INTRODUCTION TO ENGINEERING ANALYSIS

Applying Algebra II to Solve Engineering Problems

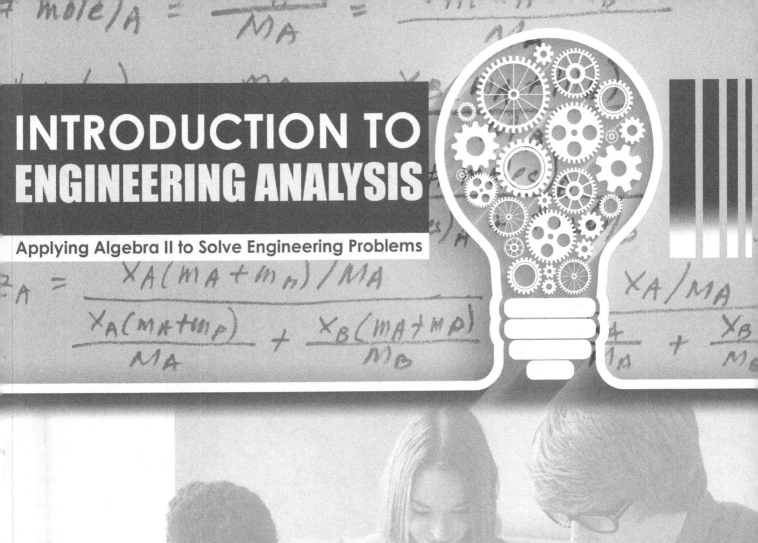

REVISED PRINTING

Edmund Tsang, Ph.D.

College of Engineering and Applied Sciences Western Michigan University

Kendall Hunt
publishing company

Cover image © Shutterstock.com

Kendall Hunt
publishing company

www.kendallhunt.com
Send all inquiries to:
4050 Westmark Drive
Dubuque, IA 52004-1840

Copyright © 2016 by Kendall Hunt Publishing Company

ISBN 978-1-4652-9050-2

Printed in the United States of America

Table of Contents

Preface

This textbook is dedicated to first-year university students who are studying engineering, engineering technology, or applied sciences and who are beginning their academic studies in Algebra II. While beginning with Algebra II put you behind the starting line of the engineering curriculum's first semester of Calculus I, this does not mean you will be unable to finish the race to obtain a bachelor's degree in engineering, engineering technology or applied sciences, and to begin your career in the profession.

My goal for the textbook is to help you, the Algebra II students, finish the race. Therefore, my objectives for writing this textbook are:

1. Demonstrate how Algebra II is applied to solve a variety of engineering problems, thus connecting you to engineering practices that use the knowledge and skills of Algebra II. I hope this will motivate you to master Algebra II now that you know how this knowledge and skills are being used, and to develop the habits of mind that you can use later in studying engineering, engineering technology, or applied sciences;

2. Provide you with additional opportunities to practice algebraic operations and manipulations, so you will gain mastery of Algebra II knowledge and skills, and develop academic habits that will be crucial to your future success. As engineers and professional artists, musicians, and athletes will attest, there is much value in practice and repetition; and

3. Allow you to develop the proper method, procedure, habit, and mindset to apply mathematics to solve problems in engineering, engineering technology or applied sciences. See Appendix I, "Instruction on Doing Engineering Homework." Once you have mastered the method, procedure, habit, and mindset, you will be able to solve increasingly more complex and more interesting engineering problems, using Pre-Calculus and Calculus, as you progress in your academic and professional career.

The chapters of this textbook are organized according to how Algebra II is taught in a 14-week course at Western Michigan University. My intention is to align, as best I can, the topics taught each week in Algebra II with applications in engineering. While I assume there is much similarity in the sequence of Algebra II topics taught across colleges in the United States, you may want to bear this in mind and follow a different chapter order.

The engineering topics that I have chosen to illustrate how Algebra II is used are by no means exhaustive, and they reflect my academic training and practice, which includes a Bachelor of Science in Mechanical Engineering, a Doctorate in Metallurgy, and post-doctoral experience in Solid State Physics.

You will notice, compared to other engineering or mathematics textbooks, I purposely included all the steps in deriving the equations and in solving the example problems. Some may consider this approach excessive. Here are the reasons why I took such a strategy in writing this textbook:

- Help you see clearly how Algebra II is used in each of the steps toward the final solution.

- Demystify the process of Algebra II operations and manipulation you use to arrive at the solution of a problem, so you can see that the process can be learned through practice and effort, if you follow a systematic approach.

- Encourage you to adopt this strategy of a systematic approach to learning.

- Through the practice and efforts of "working through the math step-by-step," I hope you will develop the proper method, procedure, habits and mindset that will help you become successful, not only during your first year of college but throughout your academic and professional career.

Once you have become proficient with the problem-solving method and with mathematics, you may skip some of the steps and still follow a systematic approach to arrive at the correct solution.

You will notice that, on occasion, I boxed a few algebraic expressions to make them stand out among the algebraic operations and manipulations.

Acknowledgment

I want to thank Dr. James Kamman, a professor of mechanical engineering (retired) and my colleague, for sharing his ideas and his help with MathType, when I was preparing the manuscript of this textbook. Dr. Kamman also reviewed my first draft of the manuscript and offered many valuable suggestions to its revisions. I am grateful for Dr. Kamman's help and friendship. Any error in the textbook is due to my oversight, and it should not be attributed to Dr. Kamman.

I also want to thank Kenneth Domingue. I have worked with Ken for the past three-and-a-half years from when he was a junior to now a first-year, master's student in aerospace engineering in teaching a course from which the materials for this textbook are based on, ENGR 1002, Introduction to Engineering Analysis. I value Ken's suggestions about the course materials from his perspective as a student and as a teaching assistant. I know the students in ENGR 1002 gave Ken great evaluation in surveys.

In addition, I want to thank Anetra Grice, Western Michigan University's College of Engineering and Applied Sciences (CEAS) STEP (STEM Talent Expansion Program) Director, for her matchless logistical supports professionally and personally as a family friend to my children Manny and Clarice Tsang that made the writing of this textbook possible and enjoyable.

I want to thank Kim Schmidt, the Kendall Hunt project coordinator assigned to me, in getting the textbook produced. Kim's assistance is always prompt and professional.

Finally, I want to acknowledge the support of the National Science Foundation in developing the course materials of ENGR 1002, which eventually resulted in this textbook. The support is via a grant from the CCLI (Course, Curriculum and Laboratory Improvement) Phase 3 program, "National Model of Engineering Mathematics Education," award #0817332.

Chapter 1
Units in Engineering Analysis

(Units and Unit Conversion)

One important difference between computations performed by engineers, engineering technologists, or applied scientists from computations performed by mathematicians is that the former often involve units.

Why should you pay attention to units in engineering analysis? I suggest the following two important reasons:

1. Avoid Disasters: In the history of engineering, there have been disasters caused by a lack of attention being paid to the units of the computations and analyses. You may have heard of the loss of a $125 million Mars orbiter by the National Aeronautics and Space Administration (NASA), as a result of NASA engineers basing their computation in metric units, while their counterparts at Lockheed Martin, a private contractor, based their work in the English or U.S. Customary units.[1] Other examples of engineering disasters resulting from a lack of attention to engineering units and unit conversion can be found at the following websites:

   * http://mentalfloss.com/article/25845/quick-6-six-unit-conversion-disasters
   * http://www.cnn.com/TECH/space/9909/30/mars.metric.02/
   * http://thehairpin.com/2011/08/metric-conversion-errors-can-be-costly/

2. Paying attention to and checking units is a quick and handy way to determine if you have made any mistakes in your analysis or computation. For example,

   * Do the units match the term you want to calculate, e.g., are the units meters or inches when you are trying to compute the length of a connecting rod?
   * Are the units on the left side of an equation equal to the units on the right side of the equation?

The *Learning Outcomes* of this chapter are:

* Demonstrate understanding of units in engineering analysis
* Apply conversion to change from S.I. to U.S. Customary units, or vice versa
* Demonstrate knowledge of $(ab)^n = a^n b^n$, and $\neq ab^n$
* Demonstrate that the magnitude or value of a parameter remains unchanged, when you multiply or divide it by one (1)

---

[1] http://www.cnn.com/TECH/space/9909/30/mars.metric.02/

1.1 Systems of Units

There are two primary systems of units used in engineering computation or analysis. Most of you have prior experience with a ruler such as shown in Figure 1.1 below.

Dmitry Guzhanin/Shutterstock.com

Figure 1.1 Ruler showing both inches and centimeters on its scales

1.1.1 International System of Units (S.I.), which is often used interchangeably with the Metric units

The base units in the S.I. system of Length, Mass, and Time are meter (m), kilogram (kg), and second(s). The symbols in the parenthesis represent the S.I. notation of the units.

Derived units are expressed in terms of the base units. Here are some examples of derived units in S.I.:

- Density = mass/volume = mass/(length)^3 = kilogram/meter^3 (kg/m^3)

- Speed = distance/time = meter/second (m/s)

- Acceleration = speed/time = $\dfrac{\left(\dfrac{meter}{sec\,ond}\right)}{sec\,ond} = \dfrac{meter}{sec\,ond^2}$ or $\dfrac{m}{s^2}$

- Force = mass × acceleration = (kilogram) × $\left(\dfrac{meter}{sec^2}\right) = kg\left(\dfrac{m}{s^2}\right) = \dfrac{kg - m}{s^2}$

  This is called a Newton (N)

- Pressure or Stress = Force/Area = $\dfrac{Newton}{meter^2}\left(\dfrac{N}{m^2}\right)$

  This called a Pascal (Pa)

1.1.2 United States Customary Units

The base units in the U.S. Customary units of length and time are expressed in foot (ft) and second(s). The symbols in the parenthesis represent the notation. The base unit for mass can be confusing for first-year students in engineering, engineering technology, or applied sciences, because we use different mass units depending on the types of problems.

The unit of mass in the U.S. Customary units for problems in solid mechanics (for example, statics or dynamics) is a *slug* (slug), and the unit of force is a *pound*. A slug works in the U.S. system like the kilogram works in the S.I. system.

$$1 \text{ N} = (1kg) x (1 \frac{m}{s^2}) \quad \text{That is, 1 Newton is required to accelerate 1 kg mass at 1 m/s}^2$$

$$1 \text{ lb} = \left(1slug\right) x \left(1 \frac{ft}{s^2}\right) \text{ That is, 1 pound (force) is required to accelerate 1 slug at 1 ft/s}^2$$

Also,

Weight (1 N) = mass × acceleration = mass (1 kg) × acceleration due to gravity (9.81 m/s²)

Weight (1 lb) = mass × acceleration = mass (1 slug) × acceleration due to gravity (32.2 ft/s²)

where 1 slug weighs 32.2 lb_m and the subscript "m" denotes mass.

In the thermal or materials sciences (for example, thermodynamics or materials science), the unit for mass is the pound (lb_m) where 1 kilogram = 2.2046 lb_m.

The derived units of density, speed, acceleration, force, and pressure or stress in the U.S. Customary units are expressed in the following:

- Density = $\frac{lb}{ft^3}$ or $\frac{lb}{in^3}$

- Speed = $\frac{ft}{s}$

- Acceleration = $\frac{ft}{s^2}$

- Force = $\frac{slug - ft}{s^2} = lb_f$

- Pressure or Stress = $\frac{lb}{in^2} (psi)$

1.2 Unit Prefixes

We use Prefixes to simplify the representation of very large or very small numbers.

1.2.1 Numbers that are Much Larger than One (1)

- Kilo = 1,000 = 10^3
- Mega = 1,000,000 = 10^6
- Giga = 1,000,000,000 = 10^9

1.2.2 Numbers that are Much Small than One (1)

- Milli = 0.001 = 10^{-3}
- Micro = 0.000,001 = 10^{-6}
- Nano = 0.000,000,001 = 10^{-9}

1.3 Unit Conversion

As the practice of engineering is a global enterprise, we will have the need to convert from S.I units to U.S. Customary units, and vice versa. We can find the conversion factors needed to convert S.I units to U.S. Customary units, and vice versa, in engineering handbooks. Google is a handy source to find the conversion factors. Here is an example of the unit conversion factor:

- Length: 1 inch (in) = 2.54 centimeter (cm)   U.S. Customary units to S.I. units

1.3.1 Key Considerations in Unit Conversation

When you want to convert S.I units to U.S. Customary units, or vice versa, you could apply the following algebraic rules:

- The numerical value or magnitude of a term remains the same when it is multiplied or divided by one (1).
- Arrange the conversion factor linking the S.I.-U.S. Customary units such that you have a magnitude of one (1). For example,

$$1 \text{ inch} = 2.54 \text{ cm} \rightarrow \frac{1in}{2.54cm} = 1, \quad \text{or} \quad \frac{2.54cm}{1in} = 1$$

Therefore, when you multiple the term you want to change the units by either $\frac{1in}{2.54cm}$ or $\frac{2.54cm}{1in}$, you would not have changed the numerical value or magnitude of the term.

- When you are changing from S.I units to U.S, Customary units, use the conversion factor having S.I. unit in the denominator so the term's unit in the numerator cancels the same unit in the denominator of the conversion factor.

1.4 Example Problems

1.4.1 Example Problem #1

Convert the speed of an automobile traveling at 30 miles per hour (mph) to feet per second (ft/sec).

From engineering handbooks, we find 1 mile (mi) = 1,760 yards (yd); 1 yard (yd) = 3 feet (ft); and 1 hour (hr) = 3,600 seconds (s).

Overview of Solution

In the problem, you have *mile* in the numerator and *hour* in the denominator in the units you want to convert from U.S. Customary to S.I. units. This means you want to arrange the conversation factors with mile in the denominator of one conversion factor and with hours in the numerator of another conversion factor such as the following:

From 1 mile = 1,760 yards $\rightarrow \dfrac{1,760\,yd}{1mi} = 1$

From 1 yard = 3 feet $\rightarrow \dfrac{3\,ft}{1yd} = 1$

From 1 hour = 3,600 seconds $\rightarrow \dfrac{1hr}{3,600s} = 1$

Therefore, 30 mph $= \dfrac{30\,mi}{1hr} \times \left(\dfrac{1,760\,yd}{1mi}\right) \times \left(\dfrac{3\,ft}{1yd}\right) \times \left(\dfrac{1hr}{3600s}\right)$

The units (*mi; hr; yd*) canceled out.

Putting the numbers in a calculator, you will get 30 $mph = 44\dfrac{ft}{s}$

1.4.2 Example Problem #2

The area of a circle is 20 cm^2. What is the area in ft^2?

From engineering handbooks, you can find that 1 in = 2.54 *cm* and 1 *ft* = 12 *in*

Overview of Solution

Since the term you want to convert is *cm*^{2}, you want to arrange the conversion factor with *cm*^{2} in the denominator, and since you want to end up in *ft*^{2}, you want to arrange the conversation factor with *ft*^{2} in the numerator. Furthermore, since area is length raised to the 2nd power (*i.e.*, length^2), you will need to raise the appropriate conversation factor to 2nd power too.

From 1 *in* = 2.54 *cm* $\rightarrow \dfrac{1in}{2.54cm} = 1$

From 1 *ft* = 12 *in* $\rightarrow \dfrac{1ft}{12in} = 1$

Area of circle = 20 *cm*^{2} = 20 *cm*$^2 \times \left(\dfrac{1in}{2.54cm}\right)^2 \times \left(\dfrac{1ft}{12in}\right)^2 = 20cm^2 \times \left(\dfrac{1in^2}{6.45cm^2}\right) \times \left(\dfrac{1ft^2}{144in^2}\right)$

The units (cm^2 and *in²*) canceled out. Using a calculator, you can determine that an area of 20 cm^2 is equal to 0.0215 *ft²*.

(Remember, $(ab)^2$ is equal to (a^2b^2), and not (ab^2). Therefore, $(2.54\ cm)^2 = 2.54^2 cm^2 = 6.45\ cm^2$.

Worksheet 1.1

a. Express $1 \times 10^{-4} \, ft^2$ in $inch^2$.

b. A cube has a volume of $1 \, m^3$. What is the volume in ft^3?

Worksheet 1.2

The density of copper is 8.94 $gram/(centimeter)^3$ or (gm/cm^3). Express the density in $pound/(inch)^3$ or (lb_m/in^3).

Given: 1 $kilogram$ = 2.2 lb_m; 1 $inch$ = 2.54 cm.

Worksheet 1.3

1. The price of gold on August 4, 2014, was \$1,291.40 per ounce (oz). A gold ring has a volume of 1.5 cm³. The density of gold is 19.32 gm/cm^3. How much did you pay for the gold ring if you purchased it on August 4, 2014? Given: 1 kilogram = 2.2 lb_m; 1 lb_m = 16 $ounce$; 1 $inch$ = 2.54 cm.

Additional Warm-Up Problems

a. Express 1 mm^2 in m^2.

b. A square has an area of 14 in^2. What is the area in cm^2?

c. Express 3458 degrees in radians.

Additional Homework Problems

1. A motor is rated at 15 hp (horsepower). What is the motor's rating in joules/hour?
   Given: 1 hp = 0.7457 kilowatt; 1 watt = 1 Joule/second; 1 hour = 3600 seconds.

2. The speed of an automobile is 0.03 kilometer/(second) or (0.03 km/s). Express the speed in miles per hour (mph). Given 1 mile = 1.609 kilometer; 1 hour = 3,600 seconds.

3. The acceleration of an automobile is 1 meter/(second)^2, or (1 m/s^2). Express the acceleration in miles/(hour)^2.

4. If (1 inch) is represented by the algebraic expression (ab),

   (a) What is a? What is b?

   (b) What is $(ab)^2$?

   (c) What is (1 inch)^2?

5. If (2.54 cm) is represented by the algebraic expression (cd).

   (a) What is c? What is d?

   (b) What is $(cd)^2$

   (c) What is (2.54cm)^2

Chapter 2
Algebraic Expression

(Density, Moles, Atomic Weight, Mass Fraction, Volume Fraction, Mole Fraction)

An algebraic expression can be thought of as "a way of representing a calculation, using letters to stand for numbers"[1]. By using letters or algebraic symbols, we can

- Convert word problems into mathematics for computation
- Allow engineers to generalize problems to handle similar conditions or situations
- Allow engineers to create a mathematical model that describe similar problems

In engineering, engineering technology or applied sciences applications, there are parameters or material properties whose definitions can be expressed as an algebraic expression. We will look at a few of them: density, a mole, atomic weight, mass fraction, volume fraction, and molar fraction.

The *Learning Outcomes* of this chapter are:

- Apply the definition of density to solve problems involving mass and volume
- Apply the definition of a mole and atomic weight to solve problems involving mass and molar fraction
- Apply the definition of mass fraction, volume fraction and molar fraction to solve problems involving mass, volume, density, and atomic weights
- Demonstrate proficiency in algebraic operations including multiplication and division; taking reciprocal; and algebraic fraction

2.1 Density

2.1.1 Definition

Density is defined as mass per unit volume. We often use the letter, ρ, to represent density; m for mass; and V for volume.

Therefore, density is $\dfrac{mass}{volume}$

---

1   McCallum, W.G.; Connally E., Hughes-Hallet D., et al. (2015). *Algebra: Form and Function, Second Edition*. Hoboken, NY: Wiley.

Using the letters, we can write density as $\dfrac{m}{V}$

Substituting ρ for density, we have $\rho = \dfrac{m}{V}$

The units for density is $\dfrac{kg}{m^3}$ or $\dfrac{gm}{cm^3}$ (kilograms per cubic meter or grams per cubic centimeter, in S.I. units), or $\dfrac{lb}{in^3}$ (pound per cubic inches, in U.S. Customary units).

2.1.2 Find Volume from Density and Mass

We can find the volume occupied by a given mass of material by looking up the material's density from an engineering handbook and by applying the following algebraic operations:

Start with definition of density $\rho = \dfrac{m}{V}$. If you want to end up with an equation containing the term you want to calculate, V, on one-side of the equation, you proceed as follows:

Multiple both sides of equation by V and cancel the V's on right-hand side equation

$$\rightarrow \quad \rho(V) = \dfrac{m}{\cancel{V}}(\cancel{V}) = m$$

Divide both sides of equation by ρ and cancel the ρ's on left-hand side equation

$$\rightarrow \dfrac{\cancel{\rho}V}{\cancel{\rho}} = \dfrac{m}{\rho}$$

Therefore,
$$V = \dfrac{m}{\rho}$$

(Checking units of right-hand side of equation: mass has units of kg while ρ has units of kg/m^3. Therefore, $\dfrac{m}{\rho}$ has units of $\dfrac{kg}{\left(\dfrac{kg}{m^3}\right)} = m^3$, which is the unit for volume on the left-hand side equation.)

2.1.3 Find Mass of Materials if Given its Volume and Density

Again, start with definition of density, $\rho = \dfrac{m}{V}$. If you want to end up with an equation containing m on one side of the equation, you proceed as follows:

Multiple both sides of equation by V and then cancel V's on right-hand side equation

$$\rightarrow \quad \rho(V) = \dfrac{m}{\cancel{V}}(\cancel{V}) = m$$

Therefore, $m = \rho V$

(Checking units of right-hand side equation: ρ has units of $\dfrac{kg}{m^3}$ and V has units of m^3. Therefore, ρV has units of kg, which is the unit for mass on left-hand side equation.)

2.2 A Mole, Atomic Weight and Number of Moles of a Material

2.2.1 Definition

A *mole* is defined as a unit of measurement used in chemistry to denote 6.022×10^{23} (known as Avagadro's number) atoms of the material.

Atomic Weight is the weight of Avagadro's number of atoms (6.022×10^{23}) of the material. The Atomic Weight of a material can be found on a Periodic Table or in engineering handbooks. We will use the alphabet, M, to denote the Atomic Weight of a material of mass, m.

The *Number of Moles* of a materials can be thought of the number of units of Avagadro's number of atoms in a material, with each unit of Avagadro's number of mass equal to the Atomic Weight. Therefore,

$$\#_{Moles} = \frac{m}{M}$$

2.3 Mass Fraction, Volume Fraction, and Molar Fraction

Composites are important engineering materials whose composition consists of two or more components in which the properties of the components can be distinct (physically or chemically distinct). When combined together, the composite can have properties that are characteristically distinctive of its components. We can design unique composite materials by manipulating the properties, the amounts, as well as the orientations of the components within each other in the composite. Examples of composite materials include concrete, fiber glass and fiber-reinforced polymers, and metal composites.

yexelA/Shutterstock.com

Figure 2.1 The texture of carbon fiber in Kevlar

Kevlar is a composite material of carbon fibers embedded in a polymer. Figure 2.1 above shows the texture and orientations of the carbon fibers in Kevlar.

The composition of a composite material can be expressed as mass fraction, volume fraction, or molar fraction of the components.

2.3.1 Mass Fraction

For a binary composite material (that is, made up of two components represented by A and B), Mass Fraction A, x_A, is defined as the mass of A divided by the total mass of the composite.

$$x_A = \frac{m_A}{m_{total}} = \frac{m_A}{(m_A + m_B)} \quad \text{where } m_A \text{ and } m_B \text{ are the masses of A and B, respectively.}$$

Similarly, Mass Fraction B, x_B, is defined as the mass of B divided by the total mass of the composite.

$$x_B = \frac{m_B}{m_{total}} = \frac{m_B}{(m_A + m_B)}$$

Therefore, $x_A + x_B = \frac{m_A}{(m_A + m_B)} + \frac{m_B}{(m_A + m_B)} = \frac{\cancel{m_A + m_B}}{\cancel{m_A + m_B}} = 1$

That is, the sum of the mass fractions of the components is equal to one (1).

You can follow the same procedure as described in *Section 2.3.1* to develop equations for mass fraction of a ternary composite composing of A, B and C, and so on.

2.3.2 Volume Fraction

For a binary composite material (that is, made up of two components represented by A and B), Volume Fraction A, y_A, is defined as the volume of A divided by the total volume of the composite.

$$y_A = \frac{V_A}{V_{total}} = \frac{V_A}{(V_A + V_B)} \quad \text{where } V_A \text{ and } V_B \text{ are the volumes of A and B, respectively.}$$

Similarly, Volume Fraction B, y_B, is defined as the volume of B divided by the total volume of the composite.

$$y_B = \frac{V_B}{V_{total}} = \frac{V_B}{(V_A + V_B)}$$

Therefore, $y_A + y_B = \frac{V_A}{(V_A + V_B)} + \frac{V_B}{(V_A + V_B)} = \left(\frac{\cancel{V_A + V_B}}{\cancel{V_A + V_B}} \right) = 1$

That is, the sum of the volume fractions of the components is equal to one (1).

You can follow the same procedure as described in *Section 2.3.2* to develop equations for volume fraction of a ternary composite composing of A, B and C, and so on.

2.3.3 Molar Fraction

For a binary composite material (that is, made up of two components represented by A and B), Molar Fraction A, z_A, is defined as the number of moles of A divided by the total number of moles of the components of the composite.

$$z_A = \frac{\#\,Moles_A}{total\,\#\,Moles} = \frac{\#\,Moles_A}{(\#\,Moles_A + \#\,Moles_B)} \quad \text{where } \#Moles_A \text{ and } \#Moles_B \text{ are the number}$$

of moles of A and B, respectively.

Recalling from 2.2.1, we can apply the equation $\#Moles = \dfrac{m}{M}$ to calculate for the number of moles of each component A and B, respectively, where m is mass and M is atomic weight.

$$z_A = \frac{\dfrac{m_A}{M_A}}{\left(\dfrac{m_A}{M_A} + \dfrac{m_B}{M_B}\right)}$$

To combine or add the two terms in parenthesis in the denominator of z_A, we need to identify a common factor that allows us to add the two terms. We do this multiplying the first term by the term (M_B/M_B), and the second term by the term (M_A/M_A).

$$z_A = \frac{\dfrac{m_A}{M_A}}{\left(\dfrac{m_A}{M_A}\left(\dfrac{M_B}{M_B}\right) + \dfrac{m_B}{M_B}\left(\dfrac{M_A}{M_A}\right)\right)} = \frac{\dfrac{m_A}{M_A}}{\left(\dfrac{m_A M_B}{M_A M_B} + \dfrac{m_B M_A}{M_A M_B}\right)} = \frac{\dfrac{m_A}{M_A}}{\left(\dfrac{m_A M_B + m_B M_A}{M_A M_B}\right)}$$

We recognize $\dfrac{1}{\left(\dfrac{m_A M_B + m_B M_A}{M_A M_B}\right)} = \dfrac{M_A M_B}{m_A M_B + m_B M_A}$

We can now write $z_A = \left(\dfrac{m_A}{\cancel{M_A}}\right)\left(\dfrac{\cancel{M_A} M_B}{m_A M_B + m_B M_A}\right) = \dfrac{m_A M_B}{m_A M_B + m_B M_A}$

(Checking units, the numerator in the right-hand side equation has units of (mass)(Atomic Weight), which are the same as the denominator. Therefore, they cancel out and the right-side equation has no units, confirming the units for fraction, which is a number, and it does not have any unit.)

In a warm-up problem, you will be asked to follow this approach and derive the equation

$$z_B = \frac{m_B M_A}{m_A M_B + m_B M_A}$$

Therefore, $z_A + z_B = \dfrac{m_A M_B}{m_A M_B + m_B M_A} + \dfrac{m_B M_A}{m_A M_B + m_B M_A} = \dfrac{\cancel{m_A M_B + m_B M_A}}{\cancel{m_A M_B + m_B M_A}} = 1$

That is, the sum of the molar fractions of the components in a composite material adds to one.

You can follow the same procedure as described in *Section 2.3.3* to develop equations for molar fraction of a ternary composite composing of A, B and C, and so on.

2.4 Density of a Composite

Consider a composite consisting of components A and B with density ρ_A and ρ_B. We will now derive the equations for the density of the A–B composite based on volume fraction or mass fraction of the components in the composite.

2.4.1 Density Based on Volume Fraction of the Components

Assume composite A–B is composed of volume fraction, y_A, of A, and volume fraction, y_B, of B. To determine the density of the composite, we start with the definition of density

$$\text{Density of composite, } \rho_{composite} = \left(\frac{mass}{volume} \right)_{composite} = \left(\frac{m_A + m_B}{V_A + V_B} \right)$$

Since we are given volume fraction, y_A, we can use its definition to find the volume of A, V_A:

$$y_A = \frac{V_A}{(V_A + V_B)} \rightarrow V_A = y_A (V_A + V_B)$$

Similarly, we can use the definition of y_B to find the volume of B, V_B:

$$y_B = \frac{V_B}{(V_A + V_B)} \rightarrow V_B = y_B (V_A + V_B)$$

To calculate the density of the composite, we need to express the masses of A and B in terms of their volumes via their respective densities.

$$\rho_A = \frac{m_A}{V_A} \rightarrow m_A = \rho_A V_A = \rho_A y_A (V_A + V_B) \text{ after we substitute the equation for } V_A \text{ from above}$$

Similarly, $\rho_B = \dfrac{m_B}{V_B} \rightarrow m_B = \rho_B V_B = \rho_B y_B (V_A + V_B)$ after we substitute the equation for V_B from above

Now we can substitute the masses and volumes of A and B into the density equation for the composite

$$\rho_{composite} = \frac{m_A + m_B}{V_A + V_B} = \frac{\rho_A y_A (V_A + V_B) + \rho_B y_B (V_A + V_B)}{y_A (V_A + V_B) + y_B (V_A + V_B)}$$

Collecting the common factor $(V_A + V_B)$ in the numerator and denominator, we have

$$\rho_{composite} = \frac{(V_A + V_B)(\rho_A y_A + \rho_B y_B)}{(V_A + V_B)(y_A + y_B)}$$

Canceling the term $(V_A + V_B)$ from numerator and denominator, and recognizing $y_A + y_B = 1$

$$\rho_{composite} = \frac{\cancel{(V_A + V_B)}(\rho_A y_A + \rho_B y_B)}{\cancel{(V_A + V_B)}(y_A + y_B)} = \frac{(\rho_A y_A + \rho_B y_B)}{\underbrace{(y_A + y_B)}_{=1}} = \rho_A y_A + \rho_B y_B$$

Density of composite of A–B is expressed in terms of volume fractions of A and B and densities of A and B.

2.4.2 Density Based on Mass Fraction of the Components

Assume composite A–B is composed of mass fraction, x_A, of A, and mass fraction, x_B, of B. To determine the density of the composite, we again start with the definition of density

$$\text{Density of composite } \rho_{composite} = \left(\frac{mass}{volume}\right)_{composite} = \left(\frac{m_A + m_B}{V_A + V_B}\right)$$

Since we are given mass fraction, we can use its definition to determine m_A and m_B from x_A and x_B:

$$x_A = \frac{m_A}{(m_A + m_B)} \rightarrow m_A = x_A(m_A + m_B)$$

Similarly, $x_B = \dfrac{m_B}{(m_A + m_B)} \rightarrow m_B = x_B(m_A + m_B)$

Now, the volumes of A and B are related to densities and masses of A and B, respectively

$$\rho_A = \frac{m_A}{V_A} \rightarrow V_A = \frac{m_A}{\rho_A} = \frac{x_A(m_A + m_B)}{\rho_A}$$

Similarly, $\rho_B = \dfrac{m_B}{V_A} \rightarrow V_B = \dfrac{m_B}{\rho_B} = \dfrac{x_B(m_A + m_B)}{\rho_B}$

Substitute the masses and volumes of A and B into the density equation for the composite

$$\rho_{composite} = \left(\frac{m_A + m_B}{V_A + V_B}\right) = \frac{x_A(m_A + m_B) + x_B(m_A + m_B)}{\left(\dfrac{x_A(m_A + m_B)}{\rho_A} + \dfrac{x_B(m_A + m_B)}{\rho_B}\right)}$$

Collect the term $(m_A + m_B)$ in both numerator and denominator of the above equation

$$\rho_{composite} = \frac{(m_A + m_B)(x_A + x_B)}{\left(m_A + m_B\right)\left(\dfrac{x_A}{\rho_A} + \dfrac{x_B}{\rho_B}\right)}$$

Canceling the $(m_A + m_B)$ in both numerator and denominator of the above equation

$$\rho_{composite} = \frac{\cancel{(m_A + m_B)}(x_A + x_B)}{\cancel{(m_A + m_B)}\left(\dfrac{x_A}{\rho_A} + \dfrac{x_B}{\rho_B}\right)} = \frac{(x_A + x_B)}{\left(\dfrac{x_A}{\rho_A} + \dfrac{x_B}{\rho_B}\right)}$$

Next, we add the two terms in the denominator, which requires us to find a common factor between the first and second term. We do this by multiplying the first term by the term $\left(\dfrac{\rho_B}{\rho_B}\right)$, which has a value of one (1) and therefore does not change the magnitude or value of the first term, and by multiplying the second term by the term $\left(\dfrac{\rho_A}{\rho_A}\right)$

$$\rho_{composite} = \frac{(x_A + x_B)}{\left(\frac{x_A}{\rho_A}\left(\frac{\rho_B}{\rho_B}\right) + \frac{x_B}{\rho_B}\left(\frac{\rho_A}{\rho_A}\right)\right)} = \frac{(x_A + x_B)}{\left(\frac{x_A\rho_B}{\rho_A\rho_B} + \frac{x_B\rho_A}{\rho_A\rho_B}\right)} = \frac{\overbrace{(x_A + x_B)}^{=1}}{\left(\frac{x_A\rho_B + x_B\rho_A}{\rho_A\rho_B}\right)}$$

Recognize $(x_A + x_B) = 1$, and $\left(\dfrac{1}{\left(\dfrac{x_A\rho_B + x_B\rho_A}{\rho_A\rho_B}\right)}\right) = \dfrac{\rho_A\rho_B}{(x_A\rho_B + x_B\rho_A)}$

Therefore, $\rho_{composite} = \dfrac{\rho_A\rho_B}{(x_A\rho_B + x_B\rho_A)}$

Density of composite A–B is expressed in terms of mass fractions of A and B and the densities of A and B.

Have you noticed that the density of a composite as a function of volume fractions, versus mass fractions, of the components yielded different equations?

You can follow the same procedure as described in *Section 2.4* to develop the density equation for a ternary composite composing of A, B, and C, and so on.

2.5 Example Problem

Due to poor casting techniques, air can get trapped in a casting, resulting in a phenomenon called porosity. Porosity describes the condition of having pocket of voids distributed in the casting. Suppose poor casting resulted in a cast aluminum cylinder to have 10% porosity by volume. You check an engineering materials handbook, and you find the density of aluminum listed as 2.699 *g/cm³*. What is the density of cast aluminum cylinder with 10% porosity by volume?

Overview of Solution

You may assume the poorly-cast aluminum cylinder as a composite composed of solid aluminum, A, and "holes," B. You are given the volume percent of holes is 10%, that is $y_B = 10\%$ or 0.1.

Let us sketch a figure for the example problem; see figure below. We are given $y_B = 10\%$ or 0.1.

"holes"—B
Solid Aluminum—A

Since you are asked to determine the density of a composite made of components A and B, let us start with the definition of density for a composite material, A and B:

$$\rho_{composite} = \left(\frac{mass}{volume}\right)_{composite} = \frac{mass_A + mass_B}{Volume_A + Volume_B} = \frac{m_A + m_B}{V_A + V_B}$$

From 2.1, you recall that mass (m) is related to volume (V) through the material property, density (ρ).

$$m_A = (\rho V)_A \text{ and } m_B = (\rho V)_B$$

From 2.3.2, volume of A is related to volume fraction of A and the total volume.

$$V_A = y_A (V_{total})$$

We assume a total volume of cast aluminum cylinder of 1 cm^3—which makes sense, since the sum of the fractions of all the components of a composite is equal to one (1) unit.

$$\rho_{composite} = \frac{(\rho_A V_A + \rho_B V_B)}{(V_A + V_B)} = \frac{(\rho_A y_A (1cm^3) + \rho_B y_B (1cm^3))}{(y_A (1cm^3) + y_B (1cm^3))} = \frac{(\rho_A y_A + \rho_B y_B)}{\underbrace{(y_A + y_B)}_{=1}}$$

$$y_A = (1 - y_B) = 1 - 0.1 = 0.9 \rightarrow V_A = 0.9 \ (1 \ cm^3) = 0.9 \ cm^3$$

The density of A (solid aluminum), ρ_A is 2.699 g/cm^3 while the density of the holes, ρ_B is 0 g/cm^3 because holes do not have any weight (no mass).

$$\rho_{composite} = \frac{\left(2.699 \frac{g}{cm^3}\right)(0.9cm^3) + \left(0 \frac{g}{cm^3}\right)(0.1cm^3)}{1cm^3} = 2.429 \frac{g}{cm^3}$$

Therefore, the density of a poorly-cast aluminum cylinder with 10% porosity is 2.429 $\frac{g}{cm^3}$

Worksheet 2.1

a. An aluminum sphere has a mass of 1 *kilogram (kg)*.

   (i) What is the volume of the aluminum sphere?

   (ii) What is the radius of the aluminum sphere? Given: Volume of a sphere = $\frac{4}{3}\pi r^3$ and Density of aluminum is 2.71 *gm/cm³*.

b. A copper-zinc alloy has 35 *gram (gm)* of copper and 65 *gm* of zinc.

   (i) What is the mass percent of copper? Mass percent of zinc?

   (ii) How many moles of copper are there in the alloy? Moles of zine?

   (iii) What is the molar percent of copper in alloy? Molar percent of zine in alloy?

Given: Atomic weight of copper = 58.93 gm/mole; atomic weight of zinc = 65.39 gm/mole

Worksheet 2.2

1. Cutting tools are often made of composites to take advantage of the hardness of some components and toughness of other components. This offers the best combination in cutting properties. Suppose a cemented carbide cutting tool used for machining contains 60 weight % tungsten carbide (WC), 20 weight % titanium carbide (TiC), 10 weight % tantalum carbide (TaC), and 10 weight % cobalt (Co). Estimate the density of the composite. Given the following densities of the composite cutting tool's components: WC = 15.77 g/cm^3; TiC = 4.94 g/cm^3; TaC = 14.5 g/cm^3; Co = 8.90 g/cm^3.

Worksheet 2.3

2. Electrical contacts are often made of composites that combine the excellent electrical conductivity of metals such as silver or copper, together with the hardness of some composites to provide wear resistance due to repeated opening and closing of the electrical contact during its lifetime. Suppose a silver (Ag)-tungsten (W) composite for an electrical contact is produced by first making a porous tungsten power metallurgy compact, then infiltrating pure silver into the pores. The density of the tungsten compact before infiltration is 15.2 g/cm^3. Calculate the volume fraction of porosity, the amount of silver that is required for infiltrating the compact, and the final weight percent of silver in the compact after infiltration. The density of pure tungsten is 19.3 g/cm^3 and pure silver is 10.49 g/cm^3. Assume the density of a pore is zero.

Additional Warm-Up Problems

a. A copper-zinc alloy has 50 grams of copper and 50 grams of zinc.
   (i) What is the volume of copper in alloy? Volume of zinc in alloy?
   (ii) What is the volume percent of copper in alloy? Volume percent of zinc in alloy?

Given: density of copper = 8.9 *gm/cm*^{3}; density of zinc = 7.13 *gm/cm*^{3}.

b. There is 1 mole of copper and 1.5 mole of zinc in a copper-zinc alloy.
   (i) What is the mass of copper in alloy? Mass of zinc in alloy?
   (ii) What is the mass percent of copper in alloy? Mass percent of zinc in alloy?
   (iii) What is the volume of copper in alloy? Volume of zinc in alloy?
   (iv) What is the volume percent of copper in alloy? Volume percent of zinc in alloy?

Additional Homework Problems

1. Borsic-reinforced aluminum containing 30 volume % boron fibers is an important high-temperature, lightweight composite material made by combining high strength of the boron fibers and the light weight of the alumni matrix. Estimate the density of the composite given the density of boron fibers is 2.36 g/cm^3 and aluminum is 2.70 g/cm^3.

2. Electrical contacts are often made of composites that combine the excellent electrical conductivity of metals such as silver or copper together with the hardness of some composite to provide wear resistance. An electrical contact material is produced by infiltrating copper (Cu) into a porous tungsten carbide (WC) compact. The density of the final composite is 12.3 g/cm^3. Assume that all of the pores are filled with copper, calculate:
   (a) The volume fraction of copper in the composite
   (b) The volume fraction of pores in the WC composite prior to infiltration, and
   (c) The original density of the WC composite before infiltration.

The density of pure Copper is 8.93 g/cm3 and Tungsten Carbide is 15.77 g/cm3.

3. An uranium sphere has a volume of 5 in^3. What is its mass in kilogram?

Given: The density of uranium is 19.05 g/cm^3.

4. A copper-nickel alloy consists of 1 mole of copper and 1.5 moles of nickel.
   (a) What is the mass of copper in the alloy?
   (b) What is the mass of nickel in the alloy?
   (c) What is the mass percent of copper in the alloy? The mass percent of nickel in the alloy?

Given: Atomic mass of copper is 63.54 g/mole; atomic mass of nickel is 58.71 g/mole.

Chapter 3
Algebraic Relations

(Relationship between Mass, Volume, and Molar Fractions)

Recall from Chapter 2 the definitions for Mass Fraction, Volume Fraction, and Molar Fraction for a composite material consisting of two components A and B:

Mass Fraction of A, $x_A = \dfrac{m_A}{m_A + m_B}$ where m_A is the mass of A and m_B is the mass of B

Mass Fraction of B, $x_B = \dfrac{m_B}{m_A + m_B}$

Volume Fraction of A, $y_A = \dfrac{V_A}{V_A + V_B}$ where V_A is the volume of A and V_B is the volume of B

Volume Fraction of B, $y_A = \dfrac{V_B}{V_A + V_B}$

Molar Fraction of A, $z_A = \dfrac{\# \, Moles_A}{\# \, Moles_A + \# \, Moles_B}$ where $\#Moles_A$ is the number of moles of A and

$\#Moles_B$ is the number of B

Molar Fraction of B, $z_B = \dfrac{\# \, Moles_B}{\# \, Moles_A + \# \, Moles_B}$

Here, $\# \, Moles_A = \dfrac{m_A}{M_A}$ and $\# \, Moles_B = \dfrac{m_B}{M_B}$

where M_A = atomic weight of A and M_B = atomic weight of B

We will now apply algebraic operations and manipulations to develop the equations expressing volume fraction in terms of mass fraction; volume fraction in terms of molar fraction; mass fraction in terms of volume fraction; mass fraction in terms of molar fraction; molar fraction in terms of mass fraction, and molar fraction in terms of volume fraction.

The Learning Outcomes for this chapter are

- Apply the definitions of mass fraction and volume fraction of a composite to derive the algebraic equations relating one to the other, and to the densities of the components of the composite.
- Apply the definition of mass fraction and molar fraction of a composite to derive the algebraic equations relating one to the other, and to the atomic weights of the components of the composite.
- Apply the definition of molar fraction and volume fraction of a composite to derive the algebraic equations relating one to the other, and to the densities and atomic weights of the components of the composite.
- Demonstrate proficiency in algebraic operations including algebraic fractions.

3.1 Relation Between Volume Fraction and Mass Fraction

Assume a composite A–B containing mass fraction of A, x_A, and mass fraction of B, x_B. We will now derive the equation for volume fractions of A and B, y_A and y_B, in terms mass fractions of A and B, x_A and x_B, and the densities of A and B, ρ_A and ρ_B.

In order to calculate volume fractions of A and B, we need to determine the volumes of A and B. However, we are given mass fractions of A and B. From mass fractions, we can write equations relating the masses of A and B to the total mass, and, through densities of A and B, we can determine the volumes of A and B and the total volume. This is how we are going to proceed.

$$x_A = \frac{m_A}{m_A + m_B} \text{ and } x_B = \frac{m_B}{m_A + m_B}$$

Multiple both left-hand and right-hand sides of above equations by $(m_A + m_B)$, you will get

$$m_A = x_A \, (m_A + m_B) \text{ and } m_B = x_B \, (m_A + m_B)$$

Mass, m, is related to volume, V, through density, ρ.

$$\rho = \frac{m}{V}$$

Multiply both sides of above equation by V and cancel V on right-hand side equation,

$$\rho\,(V) = \frac{m}{\cancel{V}}(\cancel{V})$$

Then divide both sides of equation by ρ, and cancel ρ on left-side equation:

$$\frac{\cancel{\rho}\,(V)}{\cancel{\rho}} = \frac{m}{\rho} \rightarrow V = \frac{m}{\rho}$$

Recall the definition of Volume Fraction, y,

Volume Fraction of A, $y_A = \dfrac{V_A}{V_A + V_B} = \dfrac{\dfrac{m_A}{\rho_A}}{\dfrac{m_A}{\rho_A} + \dfrac{m_B}{\rho_B}}$ by substituting $V = \dfrac{m}{\rho}$

Next, substitute $m_A = x_A (m_A + m_B)$ and $m_B = x_B (m_A + m_B)$ into the above.

Therefore, $y_A = \dfrac{\dfrac{x_A(m_A + m_B)}{\rho_A}}{\left(\dfrac{x_A(m_A + m_B)}{\rho_A} + \dfrac{x_B(m_A + m_B)}{\rho_B}\right)}$

Since $(m_A + m_B)$ is in the numerator of the above term for y_A and in both terms in the denominator, the term cancels out to give

$$y_A = \dfrac{\dfrac{x_A \cancel{(m_A + m_B)}}{\rho_A}}{\left(\dfrac{x_A \cancel{(m_A + m_B)}}{\rho_A} + \dfrac{x_B \cancel{(m_A + m_B)}}{\rho_B}\right)} = \dfrac{\dfrac{x_A}{\rho_A}}{\dfrac{x_A}{\rho_A} + \dfrac{x_B}{\rho_B}}$$

To simplify the above term, we need to combine or add the two terms in the denominator. To add the two terms in the denominator, we must find a common factor by multiplying the first term in the denominator by the term (ρ_B/ρ_B), which has a magnitude of one (1) and therefore does not change the magnitude of the first term, and by multiplying the second term by the term (ρ_A/ρ_A).

This gives, $y_A = \dfrac{\dfrac{x_A}{\rho_A}}{\dfrac{x_A}{\rho_A}\left(\dfrac{\rho_B}{\rho_B}\right) + \dfrac{x_B}{\rho_B}\left(\dfrac{\rho_A}{\rho_A}\right)} = \dfrac{\dfrac{x_A}{\rho_A}}{\dfrac{x_A\rho_B}{\rho_A\rho_B} + \dfrac{x_B\rho_A}{\rho_A\rho_B}} = \dfrac{\dfrac{x_A}{\rho_A}}{\dfrac{x_A\rho_B + x_B\rho_A}{\rho_A\rho_B}}$

Rearranging the term in the denominator, this gives us $y_A = \left(\dfrac{x_A}{\rho_A}\right)\left(\dfrac{\rho_A\rho_B}{x_A\rho_B + x_B\rho_A}\right)$

ρ_A in the numerator of the second term cancels ρ_A in the denominator of the first term, and you now have the volume fraction of A expressed in terms of mass fractions of A and B (which are given) and the densities of A and B (which you can look up the values in engineering handbooks):

$$y_A = \left(\dfrac{x_A}{\cancel{\rho_A}}\right)\left(\dfrac{\cancel{\rho_A}\rho_B}{x_A\rho_B + x_B\rho_A}\right) = \dfrac{x_A\rho_B}{x_A\rho_B + x_B\rho_A}$$

Following the same approach, you can derive the equation for volume fraction of B, y_B, in terms of mass fractions of A and B, and the densities of A and B in the Warm-Up Problems at the end of the chapter.

You can follow the same procedure as described in *Section 3.1* to develop the equations for volume fraction in terms of mass fraction for a ternary composite composing of A, B and C.

3.2 Relation Between Volume Fraction and Molar Fraction

From molar fractions of A and B, we can determine the number of moles of A and the number of moles of B. Since number of moles of A is related to the mass of A through the atomic weight of A

$$\text{\#Moles}_A = \frac{m_A}{M_A}$$ where M_A is the atomic weight of A and m_A is the mass of A.

We now have an expression for mass of A in terms of molar fractions of A and the atomic weight of A. We can then determine the volume of A from mass of A and the density of A, and hence we will be able to write the equation for volume fractions of A and B in terms of molar fractions of A and B, and the atomic weights and densities of A and B as shown below:

$$\text{Molar fraction of A, } z_A = \frac{\#\, Moles_A}{\#\, Moles_A + \#\, Moles_B}$$

$$\text{and Molar fraction of B, } z_B = \frac{\#\, Moles_B}{\#\, Moles_A + \#\, Moles_B}$$

Therefore, $\#Moles_A = z_A\,(\#Moles_A + \#Moles_B)$ and $\#Moles_B = z_B\,(\#Moles_A + \#Moles_B)$

Since $\#Moles_A = \dfrac{m_A}{M_A}$ and $\#Moles_B = \dfrac{m_B}{M_B}$ where M_A and M_B are atomic weights of A and B.

Hence, $m_A = M_A\,(\#Moles_A) = M_A z_A(\#Moles_A + \#Moles_B)$

Similarly, $m_B = M_B\,(\#Moles_B) = M_B z_B(\#Moles_A + \#Moles_B)$

Since density, $\rho = \dfrac{m}{V} \rightarrow V = \dfrac{m}{\rho}$

Therefore, $V_A = \dfrac{m_A}{\rho_A} = \dfrac{M_A z_A(\#\, Moles_A + \#\, Moles_B)}{\rho_A}$

Similarly, $V_B = \dfrac{m_B}{\rho_B} = \dfrac{M_B z_B(\#\, Moles_A + \#\, Moles_B)}{\rho_B}$

Substitute V_A and V_B into equation for volume fraction of A

$$y_A = \frac{V_A}{V_A + V_B} = \frac{\dfrac{M_A z_A\,(\#\, Moles_A + \#\, Moles_B)}{\rho_A}}{\left(\dfrac{M_A z_A\,(\#\, Moles_A + \#\, Moles_B)}{\rho_A} + \dfrac{M_B z_B\,(\#\, Moles_A + \#\, Moles_B)}{\rho_B} \right)}$$

Since the term ($\#Moles_A + \#Moles_B$) appears in the numerator of y_A as well as in both terms in the denominator, the term be cancelled out, resulting in:

$$y_A = \frac{\dfrac{M_A z_A}{\rho_A}}{\dfrac{M_A z_A}{\rho_A} + \dfrac{M_B z_B}{\rho_B}}$$

We need to simplify the above terms by combining or adding the two terms in the denominator. To do this, we must find a common factor by multiplying the first term in the denominator by the term (ρ_B/ρ_B), which has a value of one (1) and therefore does not change the magnitude of the first term, and by multiplying the second term in the denominator by the term (ρ_A/ρ_A):

$$y_A = \frac{\dfrac{M_A z_A}{\rho_A}}{\dfrac{M_A z_A}{\rho_A}\left(\dfrac{\rho_B}{\rho_B}\right) + \dfrac{M_B z_B}{\rho_B}\left(\dfrac{\rho_A}{\rho_A}\right)} = \frac{\dfrac{M_A z_A}{\rho_A}}{\left(\dfrac{M_A z_A \rho_B + M_B z_B \rho_A}{\rho_A \rho_B}\right)}$$

We recognize $\left(\dfrac{1}{\dfrac{M_A z_A \rho_B + M_B z_B \rho_A}{\rho_A \rho_B}}\right) = \dfrac{\rho_A \rho_B}{M_A z_A \rho_B + M_B z_B \rho_A}$

Rearranging the term in the denominator, we now have

$$y_A = \left(\frac{M_A z_A}{\rho_A}\right)\left(\frac{\rho_A \rho_B}{M_A z_A \rho_B + M_B z_B \rho_A}\right)$$

ρ_A in the denominator of the first term cancels ρ_A in the numerator of the second term to yield the final answer for volume fraction of A, y_A, in terms of molar fractions of A and B (which are given), and the atomic weights and densities of A and B (which you can look up in engineering handbooks):

$$y_A = \left(\frac{M_A z_A}{\not{\rho_A}}\right)\left(\frac{\not{\rho_A} \rho_B}{M_A z_A \rho_B + M_B z_B \rho_A}\right) = \frac{M_A z_A \rho_B}{M_A z_A \rho_B + M_B z_B \rho_A}$$

In one of the warm-up problems, you can follow the same approach and derive the equation for volume fraction of B, y_B, in terms of molar fractions of A and B, atomic weights of A and B, and densities of A and B.

You can follow the same procedure as described in *Section 3.2* to develop the equations for volume fraction in terms of molar fractions for a ternary composite composing of A, B and C.

3.3 Relation Between Mass Fraction and Volume Fraction

In this case, you are given the volume fractions of A and B in a composite material, and you want to derive an equation for mass fraction, which requires you to know the masses of A and B. You realize mass can be determined from volume if you know the density of the material, which you can look up in an engineering handbook. Here is how you would approach this derivation:

Volume fraction of A, $y_A = \dfrac{V_A}{V_A + V_B}$ and volume fraction of B, $y_B = \dfrac{V_B}{V_A + V_B}$

Therefore, $V_A = y_A(V_A + V_B)$ and $V_B = y_B(V_A + V_B)$

Since density, $\rho = \dfrac{m}{V} \rightarrow m = \rho V$

Therefore, $m_A = \rho_A V_A$ and $m_B = \rho_B V_B$

Substitute m_A and m_B into equations for mass fraction of A,

$$x_A = \frac{m_A}{m_A + m_B} = \frac{\rho_A V_A}{\rho_A V_A + \rho_B V_B} = \frac{\rho_A y_A (\cancel{V_A + V_B})}{\rho_A y_A (\cancel{V_A + V_B}) + \rho_B y_B (\cancel{V_A + V_B})}$$

The term $(V_A + V_B)$ is in the numerator and in both terms in the denominator, it can be canceled, yielding the following equation expressing mass fraction of A in terms of volume fractions (which are given) and densities (which you can look up in an engineering handbook):

$$x_A = \frac{\rho_A y_A}{\rho_A y_A + \rho_B y_B}$$

In a warm-up problem, you can follow the same approach and derive the equation for mass fraction of B, x_B, in terms of volume fractions and densities of A and B.

You can follow the same procedure as described in *Section 3.3* to develop the equations for mass fraction in terms of volume fractions for a ternary composite composing of A, B, and C.

3.4 Relation Between Mass Fraction and Molar Fraction

Assume a composite A–B containing molar fraction of A, z_A, and molar fraction of B, z_B. We will now derive the equation for mass fractions of A and B, x_A and x_B, in terms of the molar fractions of A and B, z_A and z_B, and the atomic weights of A and B, M_A and M_B.

In order to calculate mass fractions of A and B, we need to determine the masses of A and B. However, we are given molar fractions of A and B. From molar fractions, we can write equations relating the number of moles of A and B, and through atomic weights of A and B, to the masses of A and B. This is how we are going to proceed.

$$z_A = \frac{\#\,Moles_A}{\#\,Moles_A + \#\,Moles_B} \quad \text{and} \quad z_B = \frac{\#\,Moles_B}{\#\,Moles_A + \#\,Moles_B}$$

Multiple both left-hand and right-hand sides of the above equation by $(\#Moles_A + \#Moles_B)$ and you will get:

$$\#Moles_A = z_A\,(\#Moles_A + (\#Moles_B))$$

$$\#Moles_B = z_B\,(\#Moles_A + (\#Moles_B))$$

Since $\#\,Moles = \dfrac{Mass}{AtomicWeight} = \dfrac{m}{M}$

$\rightarrow$ Mass, $m = M\,(\#Moles)$

Therefore, $m_A = M_A(\#Moles_A) = M_A z_A\,(\#Moles_A + \#Moles_B)$

$$m_B = M_B(\#Moles_B) = M_B z_B\,(\#Moles_A + \#Moles_B)$$

Recall the definition of mass fraction, $x_A = \dfrac{m_A}{m_A + m_B}$

$$\rightarrow x_A = \frac{M_A z_A\,(\#\,Moles_A + \#\,Moles_B)}{M_A z_A\,(\#\,Moles_A + \#\,Moles_B) + M_B z_B\,(\#\,Moles_A + \#\,Moles_B)}$$

After canceling the term $(\#Moles_A + \#Moles_B)$ in the numerator and the denominator, we have

$$x_A = \frac{M_A z_A}{M_A z_A + M_B z_B}, \text{ equation for mass fraction in terms of molar fraction and atomic weight.}$$

Following the same approach, you can derive the equation for mass fraction of B, x_B, in terms of the molar fractions of A and B and the atomic weights of A and B.

In a homework problem, you can follow the approaches in *3.1* to *3.4* and derive the equations expressing molar fraction in terms of volume fraction and molar fraction in terms of mass fraction.

You can follow the same procedure as described in *Sections 3.1* to *3.4.* and develop similar equations for a ternary composite composing of A, B and C.

3.5 Example Problem

You are given the mass fraction for a ternary composite composing of A, B, and C are x_A, x_B, and x_C.

Derive the equations for molar fraction, z_A, z_B, and z_C.

Overview of Solution

In order to compute molar fraction, you need to know the number of moles of each of A, B, and C. You need mass and atomic weight to compute number of moles. You can obtain the masses of A, B, and C from the given mass fractions because the definition of mass fraction is mass divided by total mass. Therefore, you can compute the values of mass by using the given mass fraction and by assuming total mass of one (1) unit.

Mass fraction A, $x_A = \dfrac{m_A}{m_{total}} = \dfrac{m_A}{1unit}$

$\rightarrow m_A = x_A(1) = x_A$

Mass fraction B, $x_B = \dfrac{m_B}{m_{total}} = \dfrac{m_B}{1unit}$

$\rightarrow m_B = x_B$

Mass fraction C, $x_C = \dfrac{m_C}{m_{total}} = \dfrac{m_C}{1unit}$

$\rightarrow m_C = x_C$

Now, recall the definitions of #Moles and molar fraction, and write the equations for A, B, and C

Molar fraction $z_A = \dfrac{\# Moles_A}{\# Moles_{total}} = \dfrac{\# Moles_A}{\# Moles_A + \# Moles_B + \# Moles_C}$

$$= \left(\dfrac{\dfrac{x_A}{M_A}}{\dfrac{x_A}{M_A} + \dfrac{x_B}{M_B} + \dfrac{x_C}{M_C}} \right)$$

To combine or add the three terms in the denominator of the above equation, we need to identify a common factor among all three terms. We can do this by multiplying the first term by $\left(\dfrac{M_B M_C}{M_B M_C} \right)$, the second term by $\left(\dfrac{M_A M_C}{M_A M_C} \right)$, and the 3rd term by $\left(\dfrac{M_A M_B}{M_A M_B} \right)$. Because each term has a value of one (1), it would not have changed the magnitude of the terms.

Therefore, $z_A = \left(\dfrac{\dfrac{x_A}{M_A}}{\dfrac{x_A}{M_A}\left(\dfrac{M_B M_C}{M_B M_C}\right) + \dfrac{x_B}{M_B}\left(\dfrac{M_A M_C}{M_A M_C}\right) + \dfrac{x_C}{M_C}\left(\dfrac{M_A M_B}{M_A M_B}\right)} \right)$

$= \left(\dfrac{\dfrac{x_A}{M_A}}{\dfrac{x_A M_B M_C}{M_A M_B M_C} = \dfrac{x_B M_A M_C}{M_A M_B M_C} + \dfrac{x_C M_A M_B}{M_A M_B M_C}} \right)$

$= \left(\dfrac{\dfrac{x_A}{M_A}}{\dfrac{x_A M_B M_C + x_B M_A M_C + x_C M_A M_B}{M_A M_B M_C}} \right)$

Recognize that $\left(\dfrac{1}{\dfrac{x_A M_B M_C + x_B M_A M_C + x_C M_A M_B}{M_A M_B M_C}} \right) = \left(\dfrac{M_A M_B M_C}{x_A M_B M_C + x_B M_A M_C + x_C M_A M_B} \right)$

Rearrange equation to give

$z_A = \left(\dfrac{x_A}{\cancel{M_A}} \right)\left(\dfrac{\cancel{M_A} M_B M_C}{x_A M_B M_C + x_B M_A M_C + x_C M_A M_B} \right) = \left(\dfrac{x_A M_B M_C}{x_A M_B M_C + x_B M_A M_C + x_C M_A M_B} \right)$

If you follow the same procedure, you will get the molar fraction of B and C to be

$z_B = \left(\dfrac{x_B M_A M_C}{x_A M_B M_C + x_B M_A M_B + x_C M_A M_B} \right)$

$z_C = \left(\dfrac{x_C M_A M_B}{x_A M_B M_C + x_B M_A M_B + x_C M_A M_B} \right)$

Worksheet 3.1

a. Consider a composite A–B with masses m_A and m_B. If the density of A is ρ_A and the density of B is ρ_B, write the formula for the volume fraction of B, y_B in terms of m_A, m_B, ρ_A, and ρ_B.

b. Consider the same A–B composite of **#a**. If the atomic weight of A is M_A and the atomic weight of B is M_B.

  (i) Write the formula for the molar fraction of A, z_A, in terms of m_A, m_B, M_A, and M_B.

  (ii) Write the formula for the molar fraction of B, z_B, in terms of m_A, m_B, M_A, and M_B.

Worksheet 3.2

1. Consider a composite made of components A and B. Assume m_A and m_B are the mass of A and B in the composite, and V_A and V_B are the volumes of A and B in the composite. Suppose ρ_A and ρ_B are the respectively density and M_A and M_B are the respective atomic weight.

   If you are given the volume fractions of A and B in the alloy, express the mass fraction of B, x_B, in terms of the volume fraction y_A and y_B.

Worksheet 3.3

2. Consider a composite made of components A and B. Assume m_A and m_B are the mass of A and B in the composite. Suppose the atomic weights of A and B are M_A and M_B, respectively.

   If you are given the mass fractions of A and B in the alloy, express the molar fraction of A and B, z_A and z_B, in terms of the mass fraction x_A and x_B.

Additional Warm-Up Problems

a. Consider an A–B composite with volume fractions y_A and y_B. If the density of A is ρ_A and the density of B is ρ_B, and the atomic weight of A is M_A and the atomic weight of B is M_B.

Write the formula for the molar fraction of A, z_A, in terms of y_A, y_B, ρ_A, ρ_B, M_A and M_B.

b. Consider the same A–B composite of **#a**. If the atomic weight of A is M_A and the atomic weight of B is M_B. Write the formula for the molar fraction of B, z_B, in terms of y_A, y_B, ρ_A, ρ_B, M_A, and M_B.

Additional Homework Problems

a. If you are given the molar faction of A and B in the alloy, express the volume fraction of A and B, y_A and y_B, in terms of the molar fraction z_A and z_B.

b. If you are given the volume fraction of A and B in the alloy, express the molar fraction, z_A and z_B, in terms of the volume fraction, y_A and y_B.

c. A ternary alloy is made of components A, B, and C. Assume the alloy has mass fraction x_A, x_B, and x_C. If the density of components A, B, and C are ρ_A, ρ_B, and ρ_C, respectively, write the algebraic equation for volume fraction of A, B, and C in terms of the mass fraction and the density.

d. A ternary alloy is made of components D, E, and F. Assume the alloy has molar fraction z_D, z_E, and z_F. If the atomic mass of the components D, E and F are M_D, M_E, and M_F, respectively, write the algebraic equation for the mass fraction of D, E, and F in terms of the molar fraction and atomic mass.

Chapter 4
Algebraic Functions

(Coefficient of Linear Thermal Expansion)

An algebraic function is "a rule that takes numbers as inputs and assigns to each input exactly one number as the output. The output is a function of the input."[2] The input of the function is sometimes called the independent variable, and the output of the function is called the dependent variable because the value of the output *depends* on the value of the input.

We know that a metal expands when heated and contracts when cooled. In fact, the physical dimension of a metal is a function of temperature. In this chapter, we will examine the material property, coefficient of linear thermal expansion, the functional relation between geometric dimension of engineering materials and temperature, and the various types of engineering problems related to expansion and contraction.

The Learning Outcomes of this chapter are

- Apply the definition of coefficient of linear thermal expansion to solve problems involving expansion or contraction of a material as a function of temperature, or the initial or final temperature if given the initial or final dimensions.
- Demonstrate proficiency in algebraic operations to develop equations expressing the final length, initial length, final temperature, or initial temperature as a function of the other variables and the coefficient of linear thermal expansion.

4.1 Coefficient of Linear Thermal Expansion

The material property, coefficient of linear thermal expansion, is defined as the change in length per unit length, per unit change in temperature.

Change in length is given by Δl (or $l - l_o$), where l is the length at temperature T and l_o is the original length at temperature T_o. Change in length has units of m (in S.I. units) or feet (in U.S. Customary units).

Change in length per unit length is given by $\dfrac{\Delta l}{l_o}$. Change in length per unit length has no units. Ask yourself why?

Change in temperature is given by ΔT (or $T - T_o$). Change in temperature has units of $^\circ C$ (S.I. units) or $^\circ F$ (U.S. Customary units).

---

[2] McCallum, W.G., Connally, E., Hughes-Hallett, D., et al. (2015). *Algebra: Form and Function, Second Edition.* Hoboken, NY: Wiley.

Therefore, coefficient of linear thermal expansion, which is change in length per unit length, per unit change in temperature:

$$\alpha = \frac{\left(\dfrac{\Delta l}{l_o}\right)}{\Delta T} = \frac{\left(\dfrac{l-l_o}{l_o}\right)}{(T-T_o)}$$

The coefficient of linear thermal expansion has units of $^\circ C^{-1}$ (S.I. units) or $^\circ F^{-1}$ (U.S. Customary units).

4.2 Express a Material's Geometric Dimension as a Function of Temperature

Start out with the definition of the material property, coefficient of linear thermal expansion

$$\alpha = \frac{\left(\dfrac{\Delta l}{l_o}\right)}{\Delta T}$$

To end up with an equation for the final length on one-side of the equation, you proceed as follows: Multiply both sides of equation by ΔT

$$\rightarrow \alpha(\Delta T) = \frac{\left(\dfrac{\Delta l}{l_o}\right)}{\Delta T}(\Delta T) = \left(\frac{\Delta l}{l_o}\right)$$

Next, multiple both sides of equation by $l_o \rightarrow \alpha \Delta T(l_o) = \left(\dfrac{\Delta l}{l_o}\right)(l_o) = \Delta 1 = (l - l_o)$

Add l_o to both sides of equation $\rightarrow l_o + \alpha \Delta T l_o = (l - l_o) + l_o = l$

Collect the common factor l_o on left-hand-side of equation $\rightarrow l_o(1 + \alpha \Delta T) = l$

Replace ΔT by $(T - T_o)$, and the final length at temperature T is given by

$$l(T) = l_o[1 + \alpha(T - T_o)]$$

(Checking units of right-hand side equation: l_o has units of length. Inside the square bracket term, α has units of $^\circ C^{-1}$ (or $^\circ F^{-1}$) while $(T - T_o)$ has units of $^\circ C$ (or $^\circ F$); so the units cancel out each other. Therefore, the right-hand side equation has the units of length.)

4.3 Express Material's Initial Length in terms of Final Length and Temperatures

Again, start with the definition of coefficient of linear thermal expansion,

$$\alpha = \frac{\left(\dfrac{\Delta l}{l_o}\right)}{\Delta T}$$

To end up with an equation with initial length on one-side of the equation, you proceed as follows: Multiple both sides of equation by ΔT

$$\rightarrow \alpha(\Delta T) = \frac{\left(\dfrac{\Delta l}{l_o}\right)}{\Delta T}(\Delta T) = \frac{\Delta l}{l_o}$$

Multiple both sides of equation by l_o

$$\rightarrow \alpha \Delta T(l_o)\frac{\Delta l}{l_o}(l_o) = \Delta l = (l - l_o)$$

Add l_o to both sides of equation

$$\rightarrow \alpha \Delta T l_o + (l_o) = l - l_o + (l_o) = l$$

Collect the common term, l_o, on left-hand side of equation

$$\rightarrow l_o(1 + \alpha \Delta T) = l$$

Divide both sides of equation by $(1 + \alpha \Delta T)$

$$\rightarrow \frac{l_o (1 + \alpha \Delta T)}{(1 + \alpha \Delta T)} = \frac{l}{(1 + \alpha \Delta T)}$$

Therefore, initial length $l_o = \dfrac{l}{(1 + \alpha \Delta T)} = \dfrac{l}{[1 + \alpha(T - T_o)]}$

Checking units, we recognize from 4.3 that the term in the denominator of the right-hand side equation has no units, the right-hand side equation has the units of length for l.

4.4 Find the Final Temperature to Yield the Final Length

Again, start with the definition of coefficient of linear thermal expansion

$$\alpha = \frac{\left(\frac{\Delta l}{l_o}\right)}{\Delta T}$$

To end up with an equation with the final temperature on one-side of the equation, you proceed as follows: Multiple both sides of equation by ΔT

$$\rightarrow \alpha(\Delta T) = \frac{\left(\frac{\Delta l}{l_o}\right)}{\Delta T}(\Delta T) = \left(\frac{\Delta l}{l_o}\right)$$

Divide both sides of equation by α

$$\rightarrow \frac{\alpha(\Delta T)}{\alpha} = \frac{\frac{\Delta l}{l_o}}{\alpha} = \frac{\Delta l}{\alpha l_o}$$

Write ΔT as $(T - T_o)$

$$\rightarrow (T - T_o) = \frac{\Delta l}{\alpha l_o}$$

Add T_o to both sides of equation

$$\rightarrow T_o + (T - T_o) = T_o + \frac{\Delta l}{\alpha l_o}$$

Yielding final temperature, $T = T_o + \dfrac{\Delta l}{\alpha l_o}$

(Checking the units: Δl has units of length; α has units of $^oC^{-1}$ (or $^oF^{-1}$) while l_o has units of length. Therefore, the units on right-hand side equation is (length)/[$(^oC^{-1})$x(length)], or $1/(^oC^{-1})$ or ^oC.)

4.5 Find the Initial Temperature to Yield Final Length at Final Temperature

Again, start with the definition of coefficient of linear thermal expansion

$$\alpha = \frac{\left(\dfrac{\Delta l}{l_o}\right)}{\Delta T}$$

To end up with an equation with initial temperature on one side of the equation, you proceed as follows: Multiple both sides of equation by ΔT

$$\rightarrow \alpha(\Delta T) = \frac{\left(\dfrac{\Delta l}{l_o}\right)}{\Delta T}(\Delta T) = \frac{\Delta l}{l_o}$$

Divide both sides of equation by α

$$\rightarrow \frac{\alpha(\Delta T)}{\alpha} = \frac{\dfrac{\Delta l}{l_o}}{\alpha} = \frac{\Delta l}{\alpha l_o}$$

Write ΔT as $(T - T_o)$

$$\rightarrow (T - T_o) = \frac{\Delta l}{\alpha l_o}$$

Subtract T from both sides of equation

$$\rightarrow (T - T_o) - T = -T_o = \frac{\Delta l}{\alpha l_o} - T$$

Multiply both sides of equation by (-1)

$$\rightarrow (-1)(-T_o) = (-1)(\frac{\Delta l}{\alpha l_o} - T)$$

Thus yielding the initial temperature, $T_o = T - \dfrac{\Delta l}{\alpha l_o}$

(Checking the units: We have demonstrated that $\dfrac{\Delta l}{\alpha l_o}$ has units of ^oC. Therefore, $T - \dfrac{\Delta l}{\alpha l_o}$ has a unit of ^oC.)

4.6 Example Problem

A square plate has a length of l_o at 0°C. Assuming uniform expansion, what is the area of the plate at temperature T? Demonstrate that when the coefficient of linear thermal expansion, α, is small, the area can be approximated by $l_o^2(1 + 2\alpha T)$.

Overview of Solution

Start with definition of coefficient of linear thermal expansion,

$$\alpha = \frac{\left(\dfrac{\Delta l}{l_o}\right)}{\Delta T}$$

Multiple both sides of equation by ΔT and cancel ΔT on right-hand side equation

$$\rightarrow \alpha(\Delta T) = \frac{\left(\dfrac{\Delta l}{l_o}\right)}{\cancel{\Delta T}}(\cancel{\Delta T}) = \frac{\Delta l}{l_o}$$

Multiple both sides of equation by l_o and cancel l_o on right-hand side equation

$$\rightarrow \alpha \Delta T(l_o) = \frac{\Delta l}{\cancel{l_o}}(\cancel{l_o}) = \Delta l = (l - l_o)$$

Add l_o to both sides of equation and cancel l_o on right-hand side equation

$$\rightarrow \alpha \Delta T l_o + (l_o) = (l - \cancel{l_o}) + \cancel{l_o} = l$$

Collect the common term, l_o, on left-hand side of equation

$$\rightarrow l_o(1 + \alpha \Delta T) = l$$

Here, $\Delta T = T - T_o = T - 0°C = T$

Therefore, length of square plate, l, at temperature T, is $l = l_o(1 + \alpha T)$

Since area of a square is (length)^2

Therefore, the area of square plate at temperature $T = l^2 = [l_o(1 + \alpha T)]^2$

Remember from algebra, $(ab)^2 = a^2 b^2$

Area $= l_o^2(1 + \alpha T)^2 = l_o^2(1 + \alpha T + \alpha T + \alpha^2 T^2) = l_o^2(1 + 2\alpha T + \alpha^2 T^2)$

Since α is small, i.e., $\alpha \ll 1$, $\alpha^2 \ll 1$ or $\alpha^2 \ll \alpha$, and the α^2 term can be ignored.

Thus, the square plate's area at T can be approximated by $l_o^2(1 + 2\alpha T)$

Worksheet 4.1

a. Gold has a linear coefficient of thermal expansion of 14.2×10^{-6} ($^{\circ}C^{-1}$). A square gold plate has a length of 50 *cm* at 20°C. Assuming uniform expansion, what is the dimension of the gold plate at 100°C?

b. A 1025 steel wire has a length of 1.025 *m* at 20°C. If the linear coefficient of thermal expansion of 1025 steel is 12.0×10^{-6} ($^{\circ}C^{-1}$), what temperature would you need to cool the steel wire to obtain a length of 1.023 *m*?

Worksheet 4.2

1. The linear coefficient of thermal expansion for copper is 17.0×10^{-6} $^\circ C^{-1}$. A copper wire $15m$ long is cooled from $40\,^\circ C$ to $-9\,^\circ C$. How much change in length will it experience? What is the final length of the copper wire at $-9\,^\circ C$?

Worksheet 4.3

1. The radius of a piston in a piston-cylinder assembly of an automobile motor is r at room temperature (assume 20°C). What should be the minimum diameter of the cylinder if the operating temperature of the motor goes from room temperature to 1700°C?

   Note: diameter = 2 × radius

Additional Warm-Up Problems

a. A nickel ball bearing has a radius of 2.00 cm at 20°C. Assuming uniform expansion, what is the maximum temperature that the ball bearing could be exposed to if its radius is not to exceed 2.005 cm?

Given: linear coefficient of thermal expansion of nickel is 13.3×10^{-6} ($°C^{-1}$).

Additional Homework Problems

1. A square copper plate has a length, l_o, at 0°C. Assuming uniform expansion, what is the area of the copper plate at T°C? Note: Area of a square is equal to (length)^2.

2. A cubic copper plate has a length, l_o at $0°C$. Assuming uniform expansion, what is the volume of the copper plate at T°C? Note: Volume of a cube is equal to (length)^3. Assuming the coefficient of linear thermal expansion, α, is small, demonstrate the volume at T°C can be approximated by $l_0^3\{1 + 3\alpha T]$

3. A rectangle has length l and width w at 0°C. Assume the rectangle is being heated to T°C. Also assume the linear thermal expansion coefficient, α, is isotropic (that is, α has the same value along the length and width), and expansion along the length and width of the rectangle is independent of each other. What is the area of the rectangle at T°C as a function of l, w, α, and T?

4. A rectangle has length l and width w at 0°C. Assume the rectangle is being cooled to −T°C. Also assume the linear thermal expansion coefficient, α, is isotropic (that is, α has the same value along the length and width), and contraction along the length and width of the rectangle is independent of each other. What is the area of the rectangle at −T°C as a function of l, w, α, and T?

Chapter 5
Algebraic Functions

(Ohm's Law and Hooke's Law)

Two engineering examples of algebraic functions in which the concept of proportionality is introduced are Ohm's law, describing the response of a conductor or resistor circuit in response to stimulation of voltage, and Hooke's law, describing the response of a spring or system of springs in response to an externally applied force. In each case, the proportionality constant is called the equivalent resistance (in the case of Ohm's Law) or the equivalent spring constant (in the case of Hooke's Law). The equivalent resistance depends on the arrangement of resistors in the electrical circuit, and the equivalent spring constant depends on the arrangement of mechanical springs.

The Learning Outcomes of this chapter are

- Apply Ohm's law to solve some simple electrical circuit problems.
- Recognize whether resistors in circuits are arranged in series or in parallel, and be able to calculate the equivalent resistance.
- Apply Hooke's law to solve some simple problems involving mechanical springs.
- Recognize whether mechanical springs are arranged in series or in parallel, and be able to calculate the equivalent spring constant.
- Demonstrate proficiency in algebraic operations including algebraic fraction and taking reciprocal.

5.1 Ohm's Law

Consider the electric circuit as shown in Figure 5.1 below consisting of a power source (battery), a switch (which allows you to open or close the electric circuit) and a source for dissipating the power (a light bulb):

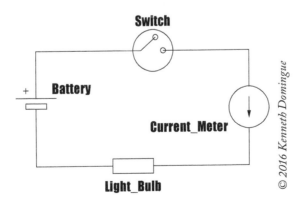

Figure 5.1 Electric Circuit

Ohm's law states that the electrical current between two points in the circuit is directly proportional to the voltage potential across the two points.

Expressed as an algebraic function, we can write Current, I, is proportional to Voltage, V,

$$I \ \alpha \ V, \text{ or}$$

$I = kV$, where I is the electric current (in units of Ampere), V is the voltage potential (in units of Volts), and k is the proportionality constant and is equal to $1/R_{eq}$ where R_{eq} is the equivalent resistance (in units of Ohms).

Re-writing the above, we have

$$V = IR_{eq}$$

Therefore, a graph of voltage, V, against current, I – with V in the vertical axis and I the horizontal axis—will be a straight line with slope equal to the equivalent resistance, R_{eq}, and the slope will pass through the origin of the graph $V = 0$, $I = 0$.

Figure 5.2 shows the backside of a circuit board consisting of many individual circuits where Ohm's Law applies.

Figure 5.2 A macro shot of backside of circuit board

5.1.1 Resistors in Series

Consider the circuit with three resistors, R_1, R_2, and R_3 arranged in series as shown in Figure 5.3:

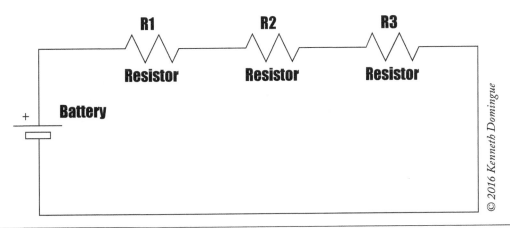

Figure 5.3 Resistors in Series

(You may think of resistors in series as customers lining up one after another in a single check-out line in a store.)

Ohm's law states $I = V/R_{eq}$

$$\text{where } R_{eq} = R_1 + R_2 + R_3$$

If there are n resistors in series

$$R_{eq} = R_1 + R_2 + \ldots\ldots + R_n$$

5.1.2 Resistors in Parallel

Consider the electrical circuit with two resistors arranged in parallel as shown in Figure 5.4 below:

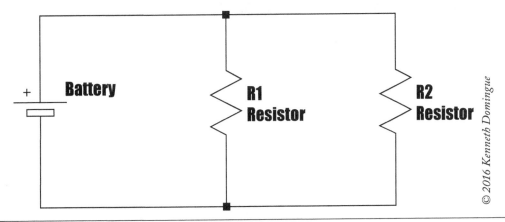

Figure 5.4 Two Resistors in Parallel

(You may think of resistors in parallel as customers lining up in two different check-out lines in a store.)

Again, Ohm's Law states $I = V/R_{eq}$

$$\text{where } \frac{1}{R_{eq}} = \frac{1}{R_1} + \frac{1}{R_2}$$

To combine or add the two terms on the right-hand side of equation, you will need to find a common factor or change the two terms to have the same denominator. To do this, multiple the first term on the right-hand side by (R_2/R_2), which has a value of one (1) and therefore does not change its magnitude, and multiply the second term by (R_1/R_1)

$$\frac{1}{R_{eq}} = \frac{1}{R_1}\left(\frac{R_2}{R_2}\right) + \frac{1}{R_2}\left(\frac{R_1}{R_1}\right) = \frac{R_2}{R_1 R_2} + \frac{R_1}{R_1 R_2} = \frac{R_1 + R_2}{R_1 R_2}$$

Taking the reciprocal of both sides of equation yields

$$R_{eq} = \frac{R_1 R_2}{(R_1 + R_2)}$$

(You can use units to check if the right-hand side equation yields the correct unit for equivalent resistance, which is ohms. The numerator of right-hand side is "ohms x ohms" and the denominator is "ohms," which yield the unit "ohms" for equivalent resistance of two resistors in parallel.)

If there are n resistors in parallel, then

$$\frac{1}{R_{eq}} = \frac{1}{R_1} + \frac{1}{R_2} + \ldots\ldots + \frac{1}{R_n}$$

5.1.3 Circuit That has Resistors in Series and in Parallel

Consider the circuit in Figure 5.5 below that has resistors in series and in parallel. Here are the rules to follow in determining the equivalent resistance:

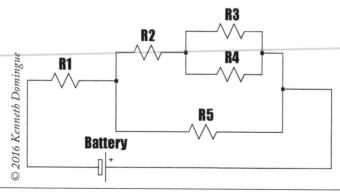

Figure 5.5 Circuit with Resistors in Series and in Parallel

In the above circuit, looking from the outside in, we see resistor R_1 is in series with another equivalent resist, which is made up of R_2, R_3, R_4 and R_5. Again, looking from outside inward, we have R_5 is parallel with R_2, which is in series with an equivalent resistor made up of R_3 and R_4. Finally, we recognize R_3 and R_4 are arranged in parallel.

Now, work from the inside of the circuit outward:

1. Combine resistors R_3 and R_4 first, which are in parallel to give R' –

$$\frac{1}{R'} = \frac{1}{R_3} + \frac{1}{R_4} = \frac{R_4}{R_3 R_4} + \frac{R_3}{R_3 R_4} = \frac{R_3 + R_4}{R_3 R_4}$$

Therefore, $R' = \dfrac{R_3 R_4}{(R_3 + R_4)}$

2. Combine resistor R_2 with equivalent resistor R', which are in series, to give R'' –

$$R'' = R' + R_2 = \frac{R_3 R_4}{(R_3 + R_4)} + R_2$$

To combine or add the two terms on right-hand side of the equation, we need to find common factors for the denominator by multiplying the second term by $(R_3+R_4)/(R_3 + R_4)$, which has a value of *one (1)* and therefore does not change its magnitude:

$$R'' = \frac{R_3 R_4}{(R_3 + R_4)} + R_2 \left(\frac{R_3 + R_4}{R_3 + R_4} \right) = \frac{R_3 R_4 + R_2 R_3 + R_2 R_4}{R_3 + R_4}$$

3. Combine resistor R_5 with equivalent R'', which are in parallel, will give R''' –

$$\frac{1}{R'''} = \frac{1}{R_5} + \frac{1}{R''} = \frac{1}{R_5} + \frac{1}{\dfrac{(R_3 R_4 + R_2 R_3 + R_2 R_4)}{R_3 + R_4}} = \frac{1}{R_5} + \frac{(R_3 + R_4)}{(R_3 R_4 + R_2 R_3 + R_2 R_4)}$$

To add or combine the two terms on right-hand side of equation, multiply the first term by $\left(\dfrac{R_3 R_4 + R_2 R_3 + R_2 R_4}{R_3 R_4 + R_2 R_3 + R_2 R_4} \right)$, which has a value of one (1) and therefore does not change its magnitude, and by multiplying the second term by $\left(\dfrac{R_5}{R_5} \right)$.

$$\frac{1}{R'''} = \frac{1}{R_5} \left(\frac{R_3 R_4 + R_2 R_3 + R_2 R_4}{R_3 R_4 + R_2 R_3 + R_2 R_4} \right) + \frac{(R_3 + R_4)}{(R_3 R_4 + R_2 R_3 + R_2 R_4)} \left(\frac{R_5}{R_5} \right) =$$

$$\frac{R_3 R_4 + R_2 R_3 + R_2 R_4}{R_5 (R_3 R_4 + R_2 R_3 + R_2 R_4)} + \frac{(R_3 + R_4) R_5}{R_5 (R_3 R_4 + R_2 R_3 + R_2 R_4)} = \frac{R_3 R_4 + R_2 R_3 + R_2 R_4 + R_3 R_5 + R_4 R_5}{R_5 (R_3 R_4 + R_2 R_3 + R_2 R_4)}$$

Taking the reciprocal of both sides of equation yields $\boxed{R'''}$

$$R''' = \frac{R_5(R_3R_4 + R_2R_3 + R_2R_4)}{R_3R_4 + R_2R_3 + R_2R_4 + R_3R_5 + R_4R_5}$$

(Again, you can check the units of the term on right-hand side of the equation to determine if it has units of "ohms" for the equivalent resistance for R'''. It does!!)

4. Finally, combine resistor R_1 and equivalent resistor R''', which are in series, to give R_{eq} –

$$R_{eq} = R_1 + \frac{R_5(R_3R_4 + R_2R_3 + R_2R_4)}{(R_3R_4 + R_2R_3 + R_2R_4 + R_3R_5 + R_4R_5)}$$

To add or combine the two terms on right-hand side equation, you need to find common factor such that the two terms have the same denominator. You do this by multiplying the first term by $\frac{(R_3R_4 + R_2R_3 + R_2R_4 + R_3R_5 + R_4R_5)}{(R_3R_4 + R_2R_3 + R_2R_4 + R_3R_5 + R_4R_5)}$, which has a value of one (1) and therefore does not change its value.

$$R_{eq} = (R_1)\left(\frac{R_3R_4 + R_2R_3 + R_2R_4 + R_3R_5 + R_4R_5}{R_3R_4 + R_2R_3 + R_2R_4 + R_3R_5 + R_4R_5}\right) + \frac{R_5(R_3R_4 + R_2R_3 + R_2R_4)}{(R_3R_4 + R_2R_3 + R_2R_4 + R_3R_5 + R_4R_5)}$$

$$R_{eq} = \frac{R_1R_3R_4 + R_1R_2R_3 + R_1R_2R_4 + R_1R_3R_5 + R_1R_4R_5 + R_5R_3R_4 + R_5R_2R_3 + R_5R_2R_4)}{(R_3R_4 + R_2R_3 + R_2R_4 + R_3R_5 + R_4R_5)}$$

(Again, check units on right-hand side equation to determine if term has units of "ohms" for R_{eq}. It does, because the numerator on right-hand side of equation has units of "ohms³" and "ohms²" in the denominator.)

5.1.4 Example Problem

Consider the following electric circuit as shown in Figure 5.6. Determine the current, I, flowing in the circuit.

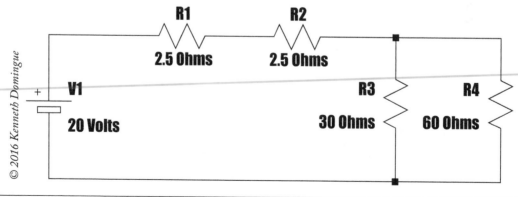

Figure 5.6 Electric Circuit for Example Problem

In this problem, resistors R_3 and R_4 are arranged in parallel, and they are arranged in series with resistors R_1 and R_2.

First, find the equivalent resistance, R', of R_3 and R_4

$$\frac{1}{R'} = \frac{1}{R_3} + \frac{1}{R_4} = \frac{1}{R_3}\left(\frac{R_4}{R_4}\right) + \frac{1}{R_4}\left(\frac{R_3}{R_3}\right) = \frac{R_3 + R_4}{R_3 R_4}$$

Therefore, $R' = \dfrac{R_3 R_4}{(R_3 + R_4)} = \dfrac{(30\Omega)(6\Omega)}{(30+6)\Omega} = 5\Omega$

Next, $\boxed{R'}$, R_1 and R_2 are in series. Therefore, the equivalent resistance,

$$R_{eq} = \boxed{R'} + R_1 + R_2 = (5 + 2.5 + 2.5)\ \Omega = 10\ \Omega$$

Ohm's Law: $V = IR_{eq}$

$$I = \frac{V}{R_{eq}} = \frac{20V}{10\Omega} = 2\,A$$

5.2 Hooke's Law

Suspension and coil springs such as those shown in Figure 5.7 below have wide engineering applications. According to Hooke's Law, the change in dimension of the springs due to a force applied along the longitude axis is proportional to the magnitude of the force.

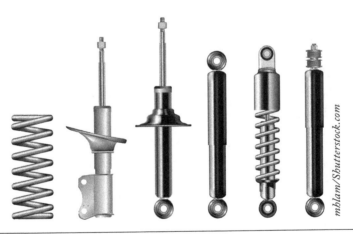

mblam/Sbutterstock.com

Figure 5.7 Suspension and coil springs used in engineering

There are similarities in the algebraic formulation of Ohm's Law and Hooke's Law of Elasticity.

Ohm's Law: $V = IR$

Hooke's Law: Consider a mechanical spring subjected to weights placed on its end as shown in Figure 5.8 below:

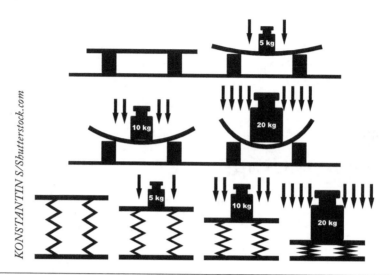

Figure 5.8 Weights Added to Spring Compressing its Length by Δy

Hooke's law states that deformation of spring is linearly proportional to the applied force due to the weight added.

$F\ \alpha\ \Delta y$ or

$F = k_{eq}\Delta y$

where F = force (in Newton or Pound)

Δy = deformation (in meter or inch)

k_{eq} = proportionality constant or equivalent spring constant (in Newton/Meter or Pound/inch)

The value of equivalent spring constant, $\boxed{k_{eq}}$, depends on the arrangements of the mechanical springs, which can be arranged in series or in parallel.

Therefore, a graph of force, F, versus displacement, Δy, will be a straight line, with slope equal to the equivalent spring constant, k_{eq}.

5.2.1 Springs Arranged in Series

The equivalent spring constant, k_{eq}, depends on the arrangement of the mechanical springs.

For example, if the springs are arranged in series as shown in Figure 5.9:

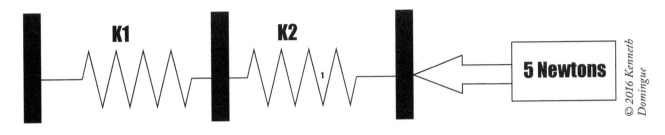

Figure 5.9 Two Mechanical Springs Arranged in Series

The equivalent spring constant, $\dfrac{1}{K_{eq}} = \dfrac{1}{K_1} + \dfrac{1}{K_2} = \dfrac{1}{K_1}\left(\dfrac{K_2}{K_2}\right) + \dfrac{1}{K_2}\left(\dfrac{K_1}{K_1}\right) = \dfrac{K_2}{K_1 K_2} + \dfrac{K_1}{K_1 K_2} = \dfrac{K_1 + K_2}{K_1 K_2}$

Taking the reciprocal of the above equation yields the equivalent spring constant for springs in series

$$K_{eq} = \frac{K_1 K_2}{(K_1 + K_2)}$$

(Again, checking the units for equivalent spring constant: The numerator in the right-hand side equation has units of (spring constant)^2 while the denominator has units of (spring constant), resulting in the right-hand side equation to have units of spring constant.)

5.2.2 Mechanical Springs Arranged in Parallel

For mechanical springs arranged in parallel as shown in Figure 5.10,

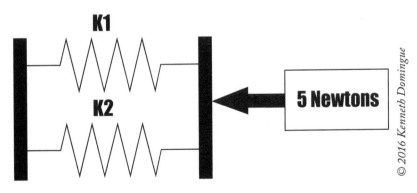

Figure 5.10 Two mechanical springs arranged in parallel

The equivalent spring constant, $K_{eq} = K_1 + K_2$

Work Sheet 5.1

a. For the electrical circuit shown below, determine the current, I.

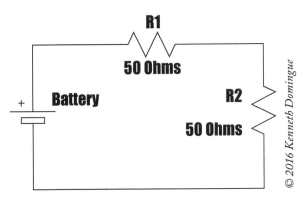

b. If R1 = 50 ohms and R2 = 75 ohms, what is the equivalent resistance for the following circuit?

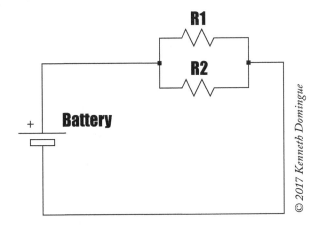

Worksheet 5.2

1. Consider an electric circuit as shown in figure below.

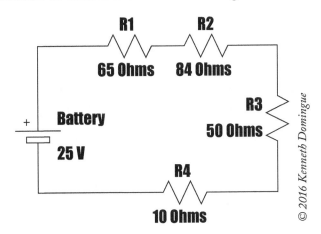

(a) What is the current, I, flowing in the circuit?

(b) What is the voltage, V, across the R_1 (65 ohm resistor)? R_2 (84 ohm resistor)? R_3 (73 ohm resistor)? R_4 (10 ohm resistor)?

(c) What is the relation between the applied voltage, 25 V, and the sum of the voltages across the four resistors?

Worksheet 5.3

1. Consider the electrical circuit below which has resistors in series as well as resistors in parallel. Determine the current, I, flowing through the circuit.

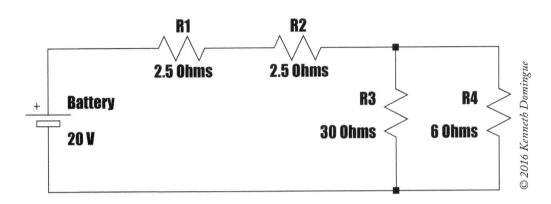

Additional Warm-Up Problem

a. Consider the following circuit. What is the value of R₂ if the equivalent resistance is 30 ohms?

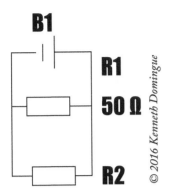

© 2016 Kenneth Domingue

b. Determine the spring constant for the follow spring:

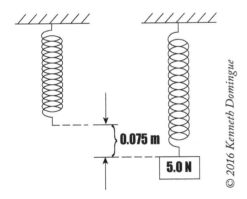

© 2016 Kenneth Domingue

Additional Homework Problems

1. An electric circuit has 3 resistors, R_1, R_2, and R_3, arranged in parallel as shown below. Write the algebraic expression for the equivalent resistance, R_{eq} in terms of R_1, R_2, and R_3. If $R_1 = 15$ ohms; $R_2 = 3$ ohms; and $R_3 = 7$ ohms. What is the numerical value of R_{eq}?

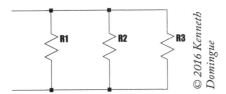

2. Consider the following circuit that has resistors in series as well as resistors in parallel.

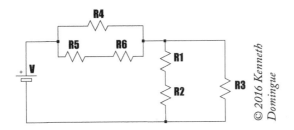

   Write the algebraic expression for R_{eq} in terms of R_1, R_2, R_3, R_4, R_5 and R_6.

3. For two springs in series, determine the deflection in centimeter for a force of 240 N and $K_1 = 120,000$ N/m and $K_2 = 80,000$ N/m.

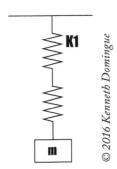

4. For the springs as shown, what force is necessary to cause a deflection of 0.015 in. Given K = 15,000 lb/in.

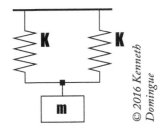

Chapter 6
Linear Functions and Equation of a Straight Line

(Linear Interpolation and Linear Extrapolation)

"Linear functions describe quantities which change at a constant rate"[1]. An equation of a straight line, graphed on a x-y Cartesian coordinate system, exemplifies linear functions with the slope of the straight line representing the constant change rate. There are many engineering examples that can be described by an equation of a straight line. In this chapter, we will apply the techniques of Linear Interpolation and Linear Extrapolation, which are based on the equation of a straight line, to approximate the value of a set of data points that are either between (interpolation) or beyond the range (extrapolation) of a known set of data points.

The Learning Outcomes for this chapter are

- Demonstrate proficiency in applying the equation of a straight line to determine slope, intercept, and the position of any point on the straight line.
- Apply the technique of linear interpolation to approximate the value of a set of data points that are in between the range of known set of data points.
- Apply the technique of linear extrapolation to approximate the value of a set of data points that are beyond the range of known set of data points.

6.1 Equation of a Straight Line

Consider two known points on a x-y Cartesian Coordinate system which are given by the coordinates (x_o, y_o) and (x_1, y_1). We draw a straight line linking (x_o, y_o) and (x_1, y_1) and extend it beyond (x_1, y_1) as shown in Figure 6.1 below:

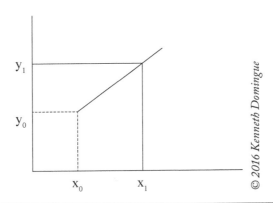

© 2016 Kenneth Domingue

Figure 6.1 Two Points (x_o, y_o) and (x_1, y_1) on x-y Cartesian Coordinates System

---

[1] McCallum, Connally, Hughes-Hallet, et al. ALGEBRA: Form and Function, Wiley, 2nd Edition, 2015.

Assume y changes linearly as x, the slope of the straight line connecting (x_o, y_o) and (x_1, y_1) is given by

$$\text{Slope, } m = \frac{Rise}{Run} = \frac{\Delta y}{\Delta x} = \frac{(y_1 - y_o)}{(x_1 - x_o)}$$

We would like to manipulate the above equation to give us y_1 as a function of x_1.

First, multiple both sides of equation by $(x_1 - x_o) \rightarrow m\left(x_1 - x_o\right) = \frac{(y_1 - y_o)}{(x_1 - x_o)} \cancel{(x_1 - x_o)} = (y_1 - y_o)$

Next, add y_o to both sides of equation $\rightarrow m(x_1 - x_o) + y_o = (y_1 - y_o) + y_o = y_1$

Therefore, $y_1 = m(x_1 - x_o) + y_o$

Assume (x_o, y_o) is the origin of the x-y coordinate system, that is $x_o = 0$ and $y_o = 0$.

If we replace y_1 by y and x_1 by x we get the general form for the equation of a straight line that passes through the origin of the x-y coordinate system

$$y = mx \quad \text{where } m \text{ is the slope of the straight line}$$

If the straight line does not pass through the coordinate's origin but intercepts the y-axis at (0, b), the equation of the straight line is

$$y = mx + b$$

6.2 Linear Interpolation

Linear Interpolation is a technique for constructing new data points *within the range* of a given discrete set of known data points. Linear interpolation is based on the assumption of a linear functional relationship between the data points.

Consider the point (x, y) between the end points (x_o, y_o) and (x_1, y_1) of the straight line as shown in Figure 6.2 below:

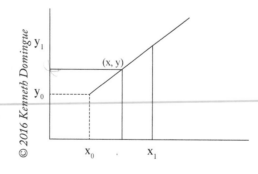

Figure 6.2

We want to write an equation for the value of y as a function of x within the range of the two end points' coordinates. We do this by writing the equation for slope, m' for the straight line between (x_1, y_1) and (x_o, y_o), and the slope of the straight line, m", between (x, y) and (x_o, y_o), and equate the slopes., $m' = m''$

$$\text{Slope, } m' = \frac{rise}{run} = \frac{\Delta y}{\Delta x} = \frac{(y_1 - y_o)}{(x_1 - x_o)}$$

$$\text{Slope, } m'' = \frac{rise}{run} = \frac{\Delta y}{\Delta x} = \frac{(y - y_o)}{(x - x_o)}$$

$$\text{Equating } m' = m'' \rightarrow \frac{(y_1 - y_o)}{(x_1 - x_o)} = \frac{(y - y_o)}{(x - x_o)}$$

We want to perform algebraic operations and manipulation to get an equation of y as a function of x.

First, multiple both sides of equation by $(x - x_o)$

$$\rightarrow \frac{(y_1 - y_o)}{(x_1 - x_o)}(x - x_o) = \frac{(y - y_o)}{(x - x_o)}(x - x_o) = y - y_o$$

Next, add y_o to both sides of equation

$$\rightarrow \frac{(y_1 - y_o)}{(x_1 - x_o)}(x - x_o) + y_o = y - y_o + y_o = y$$

Therefore, y as a function of x within the range of (x_1, y_1) and (x_o, y_o) is

$$y = y_o + \frac{(y_1 - y_o)}{(x_1 - x_o)}(x - x_o)$$

Or, $y = y_o + m(x - x_o)$ where m is slope of the straight line connecting (x_1, y_1) and (x_o, y_o).

6.3 Linear Extrapolation

Linear extrapolation is a technique for constructing new data points *outside the range* of a given discrete set of known data points. Linear extrapolation is based on the assumption of a linear relationship between the data points that extends outside the data range.

Consider the point (x, y) on the straight line outside the range defined by (x_1, y_1) and (x_o, y_o) as shown in Figure 6.3 below:

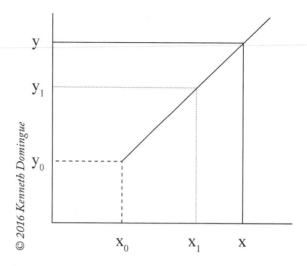

Figure 6.3 Point (x, y) is outside the range (x_o, y_o) and (x_1, y_1) on x-y axes

We will take the same approach as in 6.2 by writing, then equating, the equations for the slope of the straight line on which (x, y), (x_1, y_1) and (x_o, y_o) are data points lying on the line.

$$\text{Slope, } m' \text{ connecting } (x_o, y_o) \text{ and } (x_1, y_1) \rightarrow m' = \frac{(y_1 - y_o)}{(x_1 - x_o)}$$

$$\text{Slope, } m'' \text{ connecting } (x, y) \text{ and } (x_o, y_o) \rightarrow m'' = \frac{(y - y_o)}{(x - x_o)}$$

$$\text{Equating } m' = m'' \rightarrow \frac{(y_1 - y_o)}{(x_1 - x_o)} = \frac{(y - y_1)}{(x - x_1)}$$

We want to perform algebraic operations and manipulation to get an equation in which y is a function of x.

First, multiple both-sides of equation by $(x - x_1)$

$$\rightarrow \frac{(y_1 - y_o)}{(x_1 - x_o)}(x - x_1) = \frac{(y - y_1)}{(x - x_1)}(x - x_1) = (y - y_1)$$

Next, add y_1 to both sides of above equation and cancel y_1 on right-hand side equation

$$\rightarrow \frac{(y_1 - y_o)}{(x_1 - x_o)}(x - x_1) + (y_1) = y - y_1 + (y_1) = y$$

Therefore, $y = y_1 + \dfrac{(y_1 - y_o)}{(x_1 - x_o)}(x - x_1)$

6.4 Properties of Water

Water is an important engineering material. Water is often involved in the world of engineering as an ingredient making up a component of the engineered material, as a coolant of industrial processes (see Figure 6.4 of a cooling tower of a nuclear power plant) or is involved directly or indirectly in the generation of electricity.

Figure 6.4 The cooling towers of a nuclear power plant

Water is directly involved in the generation of electricity in a hydroelectric power plant like the Hoover Dam. Figure 6.5 shows a hydroelectric dam in the Canadian Rockies.

Figure 6.5 A dam of a hydroelectric power plant in the Canadian Rockies

Indirectly, water, in the form of super-hot and high-pressure steam (superheated steam), which is produced by the burning of a fossil or nuclear fuel, is used to drive the blades of a turbine to produce electricity. Figure 6.6 shows the steam turbine of a power generator.

arogant/Shutterstock.com

Figure 6.6 A power generator steam turbine

Therefore, the properties of water are of interest to engineers, engineering technologists, and applied scientists.

We will use the superheated steam table, an example of which is shown in Figure 6.7 below, to illustrate the concepts of Linear Interpolation and Linear Extrapolation. In the superheated steam table, the properties of water (its specific volume, specific internal energy, specific enthalpy and specific entropy) in S.I. units are given as a function of temperatures and pressures.

Table 2.6 (continued) Sample Steam Table Data

T(°C)	v(m³/kg)	u(kJ/kg)	h(kJ/kg)	s(kJ/kg · K)	v(m³/kg)	u(kJ/kg)	h(kJ/kg)	s(kJ/kg · K)
	(c) Properties of Superheated Water Vapor							
	p = 0.06 bar 0.006 MPa (Ts$_{at}$ 36.16°C)				p = 0.35 bar 0.035 MPa (Ts$_{at}$ 72.69°C)			
Sat.	23.739	2425.0	2567.4	8.3304	4.526	2473.0	2631.4	7.7158
80	27.132	2487.3	2650.1	8.5804	4.625	2483.7	2645.6	7.7564
120	30.219	2544.7	2726.0	8.7840	5.163	2542.4	2723.1	7.9644
160	33.302	2602.7	2802.5	8.9693	5.696	2601.2	2800.6	8.1519
200	36.383	2661.4	2879.7	9.1398	6.228	2660.4	2878.4	8.3237
	(d) Properties of Compressed Liquid Water							
	v × 10³				v × 10³			
	p = 25 bar = 2.5 MPa (Ts$_{at}$ 223.99°C)				p = 50 bar = 5.0 MPa (Ts$_{at}$ 263.99°C)			
20	1.0006	83.80	86.30	0.2961	0.9995	83.65	88.65	0.2956
80	1.0280	334.29	336.86	1.0737	1.0268	333.72	338.85	1.0720
140	1.0784	587.82	590.52	1.7369	1.0768	586.76	592.15	1.7343
200	1.1555	849.9	852.8	2.3294	1.1530	848.1	853.9	2.3255
Sat.	1.1973	959.1	962.1	2.5546	1.2859	1147.8	1154.2	2.9202

Source: Moran, M.J. and Shapiro, H.N. 1995. *Fundamentals of Engineering Thermodynamics*, 3rd ed. Wiley, New York, as extracted from Keenan, J.H., Keyes, F.G., Hill, P.G., and Moore, J.G. 1996. *Steam Tables*. Wiley, New York.

Figure 6.7 Superheated Steam Table taken from the CRC Handbook of Mechanical Engineering

6.5 Example Problem

The properties of Superheated Steam (S.I. Units) are shown in Figure 6.7 for temperatures from saturation to 200°C and pressures from 0.006 MPa to 0.035 MPa. Determine the specific volume, v, at $P = 0.025$ MPa and $T = 80$°C.

Overview of Solution

After reviewing the superheated steam table, first of all we recognize that there are no tabulated values for $P = 0.025$ MPa. The second thing we notice about the table is that the pressure, $P = 0.025$ MPa is between the range $P = 0.006$ MPa and $P = 0.035$ MPa for the temperature $T = 80$°C. To solve this example problem, we will need to perform linear interpolation between the pressures 0.006 MPa and 0.035 MPa to obtain the specific volume for 0.025 MPa.

From Figure 6.7, we read the specific volume of superheated steam at 80°C and 0.006 MPa and get 27.132 m³/kg, and 80°C and 0.035 MPa and get 4.625 m³/kg.

We can write the equation of v as a function of P, and the rate of change or slope of v vs. P:

Between 0.006 MPa and 0.035 MPa: slope, $m' = \dfrac{(4.625 - 27.132)}{(0.035 - 0.006)} \dfrac{m^3/kg}{MPa} = -776.103 \dfrac{m^3}{kg - MPa}$

Let Y be the value of specific volume, v, at 0.025 MPa. Between 0.006 and 0.025 MPa.

Slope, $m'' = \dfrac{(Y - 27.132)}{(0.025 - 0.006)} \dfrac{m^3/kg}{MPa}$

Equating slopes and solving for

$Y = 27.132 \dfrac{m^3}{kg} + ((0.019\ \cancel{MPa})(-776.103 \dfrac{m^3}{kg - \cancel{MPa}})) = 12.386 \dfrac{m^3}{kg}$ at $P = 0.025$ MPa and 80°C.

Worksheet 6.1

a. What is the x-y coordinate of the mid-point between (2, 3) and (10, 5), assuming a linear relation between the two coordinate points.

b. Consider the figure below. What is the y value for the x value of 5?

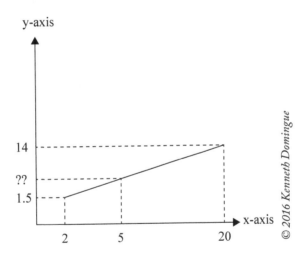

Worksheet 6.2

Refer to the following Superheated Steam Table for this homework assignment. Do not use the interpolation or extrapolation function of your calculator, but apply the algebraic skills in this homework assignment. Show all the steps in your calculation.

Table 2.6 (continued) Sample Steam Table Data

T(°C)	v(m³/kg)	u(kJ/kg)	h(kJ/kg)	s(kJ/kg · K)	v(m³/kg)	u(kJ/kg)	h(kJ/kg)	s(kJ/kg · K)
	p = 0.06 bar 0.006 MPa (Ts$_{at}$ 36.16°C)				p = 0.35 bar 0.035 MPa (Ts$_{at}$ 72.69°C)			
Sat.	23.739	2425.0	2567.4	8.3304	4.526	2473.0	2631.4	7.7158
80	27.132	2487.3	2650.1	8.5804	4.625	2483.7	2645.6	7.7564
120	30.219	2544.7	2726.0	8.7840	5.163	2542.4	2723.1	7.9644
160	33.302	2602.7	2802.5	8.9693	5.696	2601.2	2800.6	8.1519
200	36.383	2661.4	2879.7	9.1398	6.228	2660.4	2878.4	8.3237

(c) Properties of Superheated Water Vapor

Republished with the permission of Wiley from Steam Tables: Thermodynamic Properties of Water by Keenan, Keyes, Hill, and Moore. © 1969 Wiley, New York; permission conveyed through Copyright Clearance Center, Inc.

Give the numerical value of the internal energy, u, of superheated water vapor at a temperature of 155°C and a pressure of 0.035 MPa.

Worksheet 6.3

Refer to the following Compressed Liquid Water Table for this homework assignment. Do not use the interpolation or extrapolation function of your calculator, but apply the algebraic skills in this homework assignment. Show all the steps in your calculation.

Table 2.6 (continued) Sample Steam Table Data								
	(d) Properties of Compressed Liquid Water							
	$v \times 10^3$				$v \times 10^3$			
T(°C)	v(m³/kg)	u(kJ/kg)	h(kJ/kg)	s(kJ/kg · K)	v(m³/kg)	u(kJ/kg)	h(kJ/kg)	s(kJ/kg · K)
	p = 25 bar = 2.5 MPa (Ts_{at} 223.99°C)				p = 50 bar = 5.0 MPa (Ts_{at} 263.99°C)			
20	1.0006	83.80	86.30	0.2961	0.9995	83.65	88.65	0.2956
80	1.0280	334.29	336.86	1.0737	1.0268	333.72	338.85	1.0720
140	1.0784	587.82	590.52	1.7369	1.0768	586.76	592.15	1.7343
200	1.1555	849.9	852.8	2.3294	1.1530	848.1	853.9	2.3255
Sat.	1.1973	959.1	962.1	2.5546	1.2859	1147.8	1154.2	2.9202

Republished with the permission of Wiley from Steam Tables: Thermodynamic Properties of Water by Keenan, Keyes, Hill, and Moore. © 1969 Wiley, New York; permission conveyed through Copyright Clearance Center, Inc.

Give the numerical value of the internal energy, u, of compressed liquid water at a temperature of 120°C and a pressure of 3.0 MPa. (Hint: you need to do "double" linear interpolations; first at the given temperature/pressure, and then at the given pressure/temperature.)

Additional Warm-Up Problems

a. For Problem #a in Worksheet 6.1, what is the x value for a y value of 10.25?

b. For Problem #b in Worksheet 6.1, what is the y value if x = 20.5?

Additional Homework Problems

Refer to the following Superheated Steam and Compressed Liquid Water Table for this homework assignment. Do not use the interpolation or extrapolation function of your calculator, but apply the algebraic skills in this homework assignment. Show all the steps in your calculation.

Table 2.6 (continued) Sample Steam Table Data

(c) Properties of Superheated Water Vapor

T(°C)	v(m³/kg)	u(kJ/kg)	h(kJ/kg)	s(kJ/kg · K)	v(m³/kg)	u(kJ/kg)	h(kJ/kg)	s(kJ/kg · K)
	p = 0.06 bar 0.006 MPa (Ts_{at} 36.16°C)				p = 0.35 bar 0.035 MPa (Ts_{at} 72.69°C)			
Sat.	23.739	2425.0	2567.4	8.3304	4.526	2473.0	2631.4	7.7158
80	27.132	2487.3	2650.1	8.5804	4.625	2483.7	2645.6	7.7564
120	30.219	2544.7	2726.0	8.7840	5.163	2542.4	2723.1	7.9644
160	33.302	2602.7	2802.5	8.9693	5.696	2601.2	2800.6	8.1519
200	36.383	2661.4	2879.7	9.1398	6.228	2660.4	2878.4	8.3237

(d) Properties of Compressed Liquid Water

T(°C)	$v \times 10^3$ v(m³/kg)	u(kJ/kg)	h(kJ/kg)	s(kJ/kg · K)	$v \times 10^3$ v(m³/kg)	u(kJ/kg)	h(kJ/kg)	s(kJ/kg · K)
	p = 25 bar = 2.5 MPa (Ts_{at} 223.99°C)				p = 50 bar = 5.0 MPa (Ts_{at} 263.99°C)			
20	1.0006	83.80	86.30	0.2961	0.9995	83.65	88.65	0.2956
80	1.0280	334.29	336.86	1.0737	1.0268	333.72	338.85	1.0720
140	1.0784	587.82	590.52	1.7369	1.0768	586.76	592.15	1.7343
200	1.1555	849.9	852.8	2.3294	1.1530	848.1	853.9	2.3255
Sat.	1.1973	959.1	962.1	2.5546	1.2859	1147.8	1154.2	2.9202

Source: Moran, M.J. and Shapiro, H.N. 1995. *Fundamentals of Engineering Thermodynamics*, 3rd ed. Wiley, New York, as extracted from Keenan, J.H., Keyes, F.G., Hill, P.G., and Moore, J.G. 1996. *Steam Tables*. Wiley, New York.

Republished with the permission of Wiley from Steam Tables: Thermodynamic Properties of Water by Keenan, Keyes, Hill, and Moore. © 1969 Wiley, New York; permission conveyed through Copyright Clearance Center, Inc.

1. Give the numerical value of the specific internal energy, u, of superheated water vapor at a temperature of 140°C and a pressure of 0.010 MPa.

2. The specific entropy, s, of compressed liquid water at a temperature of 140°C is 1.7355 kJ/kg-K. What is the pressure in MPa?

3. Give the numerical value of the specific enthalpy, h, of superheated water vapor at a temperature of 205°C and a pressure of 0.006 MPa.

4. Give the numerical value of the specific volume, v, of compressed liquid water at 200°C and 5.50 MPa.

5. The specific enthalpy, h, of superheated steam at 0.035 MPa is 2900 kJ/kg. What is the T?

Chapter 7
Equation of a Straight Line

(Position, Speed, and Acceleration)

Another example of applying a linear function to describe engineering problems is that of position and constant motion—more specifically, position and constant speed or constant acceleration. Knowing the initial conditions, such as initial position or initial speed, we can describe completely the position or speed at a later time, when given the value of the constant speed or constant acceleration.

The Learning Outcomes of this chapter are

- Apply the definition of constant speed to calculate the position at any future time in terms of the initial position and the value of the constant speed.
- Apply the definition of constant acceleration to calculate the speed at any future time in terms of the initial speed and the value of the constant acceleration.
- Demonstrate proficiency in applying the equation of a straight line to determine slope, intercept, and the position of any point on the straight line.

7.1 Equation of a Straight Line

Recall the equation of a straight line is

$$y = mx + b$$

where m = slope; b = intercept; x is the input; and y is the output of the linear function

$$m = \frac{rise}{run} = \frac{\Delta y}{\Delta x} = \frac{(y - y_o)}{(x - x_o)}$$

Perform algebraic operations and manipulation to obtain an equation of y as a function of x:

First, multiple both sides of above equation by $(x - x_o)$

$$\rightarrow m(x - x_o) = \frac{(y - y_o)}{\cancel{(x - x_o)}} \cancel{(x - x_o)} = (y - y_o)$$

Add y_o to both sides of above equation

$$\rightarrow m(x - x_o) + \left(y_o \right) = y - \cancel{y_o} + \cancel{y_o}$$

Therefore, $y = y_o + m\,(x - x_o)$ is the equation of a straight line, where $b = y_o$ is the y-intercept.

111

7.2 Equations Relating Position, Time, Speed and Acceleration

The following equations are developed to relate position with time, speed, and acceleration of a particle:

7.2.1 Constant Speed

Speed, V, is defined as change in position, S, as a function of time, t.

$$\text{Speed, } V = \frac{\Delta S}{\Delta t} = \frac{(S_1 - S_o)}{(t_1 - t_o)}$$

where S_1 is position of particle at time t_1 and S_o is the position of particle at time t_o.

For constant speed, V, the rate of change of position as a function of time is a constant, and position S is a linear function of time t.

7.2.1.1 Equation of Position as a Function of Time for Constant Speed

Start with the equation of constant speed, and perform the algebraic operations and manipulation to write an equation for the position (S) of particle as a function of time (t) and the initial condition of time and position of particle, (t_o, S_o) as follows:

First, multiply both sides of speed equation by $(t_1 - t_o)$

$$\rightarrow V(t_1 - t_o) = \frac{(S_1 - S_o)}{(t_1 - t_o)} (t_1 - t_o) = (S_1 - S_o)$$

Add S_o to both sides of above equation

$$\rightarrow V(t_1 - t_o) + (S_o) = (S_1 - S_o) + (S_o) = S_1$$

Rearranging the equation gives

$$S_1 = S_o + V(t_1 - t_o)$$

Or, $S(t) = S_o + V(t - t_o)$, which gives the equation for position S as a function of time *(t)*, *the* constant speed (V), and the initial position *(S_o)*.

7.2.1.2 Determine the Initial Position

From constant speed equation, we can determine the initial condition (initial position, S_o) at the start of the change from the position at a later time S_1 at time t_1.

Start out with constant speed V equation: $V = \frac{\Delta S}{\Delta t} = \frac{(S_1 - S_o)}{(t_1 - t_o)}$

Multiple both sides of equation by $(t_1 - t_o)$

$$\to V(t_1 - t_o) = \frac{(S_1 - S_o)}{(t_1 - t_o)}(t_1 - t_o) = S_1 - S_o$$

Subtract S_1 from both sides of above equation

$$\to V(t_1 - t_o) - (S_1) = S_1 - S_o - (S_1) = -S_o$$

Multiple both sides of equation by "-1"

$$\to -V(t_1 - t_o) + S_1 = S_o \text{ or } S_o = S_1 - V(t_1 - t_o)$$

7.2.2 Constant Acceleration

Acceleration is defined as change in speed, V, as a function of time.

$$\text{Acceleration, } a = \frac{\Delta V}{\Delta t} = \frac{(V_1 - V_o)}{(t_1 - t_o)}$$

where V_1 is the speed at time t_1, and V_O is the speed at time t_O.

For constant acceleration, the rate of change of speed, V, as a function of time, t, is a constant; and speed, V, is a linear function of time, t.

7.2.2.1 Equation of Speed As a Function of Time for Constant Acceleration

Start with the equation of constant acceleration, and perform the algebraic operations and manipulation to write an equation for the speed (V) of particle as a function of time (t) and the initial condition of speed of particle V_o at time t_o as follows:

$$a = \frac{(V_1 - V_o)}{(t_1 - t_o)}$$

First, multiply both sides of speed equation by $(t_1 - t_O)$

$$a(t_1 - t_o) = \frac{(V_1 - V_o)}{(t_1 - t_o)}(t_1 - t_o) = V_1 - V_o$$

Add V_O to both sides of above equation

$$\to a(t_1 - t_o) + (V_o) = V_1 - V_o + (V_o) = V_1$$

Rearranging the equation gives

$$V_1 = V_o + a(t_1 - t_o)$$

Or, $V(t) = V_o + a(t - t_o)$, which gives the equation for speed as a function of time *(t)*, *at* constant

acceleration (*a*), and the initial speed V_o.

7.2.2.2 Determine the Initial Speed

From constant acceleration, we can determine the initial condition (initial speed, V_o) at the start of the change, from the Speed at a later time, V_1 at time t_1.

Start out with constant acceleration equation,

$$a = \frac{\Delta V}{\Delta t} = \frac{(V_1 - V_o)}{(t_1 - t_o)}$$

We perform algebraic operations and manipulation to write an equation for V_o as a function of time and the speed at a later time V_1 at time t_1.

Start out by multiplying both sides of equation by $(t_1 - t_o)$

$$\rightarrow a(t_1 - t_o) = \frac{(V_1 - V_o)}{(t_1 - t_o)} (t_1 - t_o) = V_1 - V_o$$

Subtract V_1 from both sides of equation

$$\rightarrow a(t_1 - t_o) - (V_1) = V_1 - V_o - (V_1) = -V_o$$

Multiple both sides of above equation by "–1"

$$\rightarrow -a(t_1 - t_o) + V_1 = V_o \text{ or } V_o = V_1 - a(t_1 - t_o)$$

7.3 Example Problem

Consider a moving vehicle that is subjected to a constant braking force, resulting in constant deceleration. The velocity of the vehicle during braking has been measured at two distinct points in time as shown in Table 1 below. Determine

a) The deceleration
b) The speed just prior to braking
c) The time required for vehicle to stop completely
d) Graph the equations representing the velocity during braking as a function of time

Table 7.1 Vehicle Speed During Breaking as a Function of Time

t (s)	V (t) (m/s)
1.5	9.75
2.5	5.85

Overview of Solution

Using the provided data, you can determine the deceleration as $\Delta V/\Delta t$. Then, following 7.2.2.2, we can determine the speed V_o right before deceleration took place; equating $V(t) = 0$, we can determine the time when speed is zero; and graph $V(t)$.

a) The deceleration

$$a = \frac{\Delta V}{\Delta t} = \frac{(V_1 - V_o)}{(t_1 - t_o)} = \frac{(5.85 - 9.85)\ m/s}{(2.5 - 1.5)\ s} = -4.0\frac{m}{s^2}$$

b) From 7.2.2.2, we have $V_o = V_1 - a(t_1 - t_o)$ in which $t_o = 0$ sec

$$V_o = 9.85(m/s) - (-4\frac{m}{s^2})(1.5 - 0)s = 15.85 \text{ m/s}$$

which is speed when the brake was applied

c) From 7.2.2.1, we have $V(t) = V_o + a(t - t_o)$

When vehicle stopped completely, $V(t) = 0$ m/s. Therefore, by equating the equation for $V(t) = 0$, we can solve for t when $V(t)$ is 0 m/s.

$$0 = V_o + a(t - t_o) = 15.85 \text{ m/s} + (-4 \text{ m/s}^2)(t - 0)s$$

Or $4t = 15.85$ or $t = 3.961$ secs

Therefore, it takes 3.96 s from the moment brake was applied to time when car stopped completely.

Worksheet 7.1

a. An automobile traveling at constant speed is at distance marker 75 kilometer (km) at 2:05 p.m. and 86.5 km at 2:16 p.m. What is the speed of the automobile in mph (miles per hour)?

b. An automobile is traveling at constant speed along a straight line. It traveled 250 feet in 20 seconds. If the automobile continues the same speed along the straight line, what distance would be traveled in 45 seconds?

Worksheet 7.2

1. A particle traveling at constant speed along a straight line passes the position $S = 10$ cm at time $t = 1.5$ second and the position $S = 20$ cm at time $t = 5.0$ second.

   (a) What is the speed of the particle?

   (b) What is the position of the particle at $t = 0$ second?

   (c) What is the position of the particle at $t = 10$ seconds?

   (d) Write the equation for position as a function of time, $S(t)$.

Worksheet 7.3

2. A particle traveling at a constant speed of 20 ft/s passes the position $S = 500$ feet at time $t = 12$ seconds. Assume the relationship between position, S, and time, t, is given by $S(t) = S_o + Vt$.

(a) Determine the starting position, S_o, at time, $t = 0$ s.

(b) Sketch of graph of $S(t)$ and clearly label the initial position and speed.

(c) What time does the participle reach the position $S = 1000$ feet?

Additional Warm-Up Problems

a. A particle traveling at 5 cm/s is given a constant acceleration of 7.5 cm/s^2.

   (i) How long would it take the particular to reach a speed of 25 cm/s?

   (ii) What is the particle's speed after 15 seconds at the same constant acceleration?

b. An automobile is traveling at 65 mph when the driver presses the accelerator to give it a constant acceleration. The automobile reaches 70 mph after 15 seconds. What is the acceleration in $feet/s^2$? What is the speed of the automobile 5 seconds after it reaches 70 mph?

Additional Homework Problems

1. A vehicle starts from a resting position and accelerates at a constant value of 4 ft/s^2 (feet/second²). How long will it take the vehicle to reach a speed of 70 mph (miles per hour)? Given: 1 mile = 5,280 feet.

2. The speed of a vehicle is measured at two distinct points in time as given by the table below. The speed satisfies the relationship, $V(t) = V_o + at$, where V_o is the initial velocity in meter/second (m/s) and a is the acceleration in m/s^2.

$V(t)$, m/s	30	10
t, s	1.0	2.0

   a. Find the equation for $V(t)$ and determine both the initial speed, V_o and the acceleration, a.

   b. Sketch the graph of $V(t)$, and clearly labeled the initial speed and the acceleration, a.

   c. Determine the stopping time, that is, the time it takes for vehicle comes to a stop.

3. The velocity of a vehicle is given as a function of time as shown

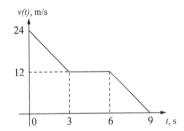

   (a) Determine the acceleration, a, for the time intervals, t, between 0 to 9 seconds.

   (b) Plot the acceleration of the vehicle from $t = 0$ second to $t = 9$ seconds.

   (c) Write the equation for speed, $V(t)$, for the time intervals, t, between 0 to 9 seconds.

4. A particle starts at rest (i.e., speed = 0 m/s) is subjected to an acceleration of 2 m/s^2 for 5 seconds, followed by a deceleration of 1 m/s^2 for the next 5 seconds.

   (a) Sketch the acceleration profile from 0 to 10 seconds;

   (b) Write the equation for speed, from 0 to 10 seconds;

   (c) Sketch the speed profile from 0 to 10 seconds.

5. A particle starts at the origin of the x-axis and travels at a constant speed of 2.5 ft/s for 10 seconds, followed by a constant speed of 3.0 ft/s for the next 5 seconds.

   (a) Sketch the speed profile from 0 to 15 seconds;

   (b) Write the position equation, from 0 to 15 seconds;

   (c) Sketch the position profile from 0 to 15 seconds.

Chapter 8
Quadratic Equations

(Projectiles)

A quadratic function in standard form can be represented by

$$y = f(x) = ax^2 + bx + c$$

where *a, b, c* are constants, and $a \neq 0$.

In this chapter, we will use the Quadratic Formula to give solutions to the equation $ax^2 + bx + c = 0$.

The Quadratic Formula says $x = \dfrac{-b \pm \sqrt{b^2 - 4ac}}{2a}$

The Learning Outcomes for this chapter are

- Apply the quadratic equation describing the path of a projectile to determine the time of flight at any height.
- Be able to interpret the positive and negative values that resulted from the solutions to a quadratic equation in projectile problems.
- Demonstrate proficiency in using the Quadratic Formula in solving projectile problems.

8.1 Projectiles in Engineering

Projectile problems in engineering involving constant acceleration can be described by quadratic functions. For example, the height of a ball thrown upward from ground can be described as a quadratic function of time, *t,*

$$h(t) = at^2 + bt + c$$

where *a, b,* and *c* are constants, and $a \neq 0$

Under certain conditions, the flight of a golf ball in horizontal and vertical axes can be expressed as function of time, *t,* as follows:

$$y(t) = at^2 + bt + c$$

$$x(t) = dt$$

where *a, b, c, and d* are constants, and $a \neq 0$.

The path of a projectile such as shown in Figure 8.1 below can be described by a quadratic equation as a function of time. The red line in the Picture can be described by a quadratic equation, $y(x) = px^2 + qx + r$ where p, q, and r are constants, $p \neq 0$.

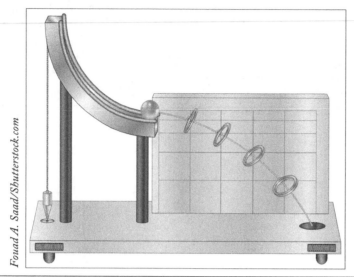

Figure 8.1 The path of a horizontally fired projectile

8.2 Example Problem

A model rocket is launched in the vertical plane at time $t = 0$ second as shown in Figure 8.2 below.

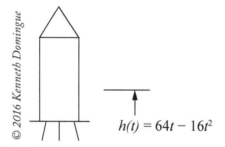

Figure 8.2 Model rocket is Launched from Ground at $t = 0$ sec.

The height of the rocket, $h(t)$, satisfied the quadratic equation

$$h(t) = 64t - 16\,t^2$$

a. Find the times, t, when $h(t) = 48$ ft, 60 ft
b. Find the time, t, required for the rocket to hit the ground.
c. Determine the maximum height reached by the rocket.

Overview of Solution

Since the height of rocket is a quadratic function of time, a graph of height versus time is a parabola. There will be two times, t, for every height value—one for the rocket ascending and the other for the rocket descending. When the rocket hits the ground, the height will be 0 feet. There will be only one time at which the rocket reaches the maximum height, and the maximum height is reached at mid-flight.

a. $h(t) = 48$ ft $= 64t - 16t^2$

Rewriting quadratic equation as $16t^2 - 64t + 48 = 0$

Compared to the Standard Form of $ax^2 + bx + c = 0$

$$\rightarrow a = 16; \; b = -64; \; \text{and } c = 48; \; \text{and } x = t$$

Apply the Quadratic Formula to solve the quadratic equation

$$x = \frac{-b \pm \sqrt{b^2 - 4ac}}{2a}$$

Therefore, $t = \dfrac{-(-64) \pm \sqrt{(-64)^2 - 4(16)(48)}}{2(16)} = \dfrac{64 \pm 32}{32}$ s

Time, t, is 1 s or 3 s into flight when height is at 48 feet.

$h(t) = 60$ ft $= 64t - 16t^2$

Rewriting quadratic equation as $16t^2 - 64t + 60 = 0$ and compared to Standard Form

$$\rightarrow a = 16, \; b = -64, \; c = 60$$

Apply Quadratic Formula to solve for t

$$t = \frac{-(-64) \pm \sqrt{(-64)^2 - 4(16)(60)}}{2(16)} = \frac{64 \pm (16)}{32} \text{ s}$$

Therefore, t is 1.5 s and 2.5 s into flight when rocket reaches 60 ft

b. $h(t) = 0$ ft when rocket hits the ground

Therefore, $0 = 64t - 16t^2$

Rewriting quadratic equation as $16t2 - 64t + 0 = 0$ and compared to Standard Form

$$\rightarrow a = 16, \; b = -64, \; \text{and } c = 0$$

Apply Quadratic Formula to solve for t

$$t = \frac{-(-64) \pm \sqrt{(-64)^2 - 4(16)(0)}}{2(16)} = \frac{64 \pm 64}{32}$$

Therefore, $t = 0$ s and 4 s when rocket hits ground

c. Rocket is at maximum height at mid-flight

$$\rightarrow t_{mid-flight} = \frac{4s}{2} = 2s$$

$h(2\ s) = 64(2) - 16(2^2) = 64\ ft$

Worksheet 8.1

a. A model rocket is fired into the air from the ground. The height of the rocket, in meters, satisfied the equation $h(t) = 98t - 4.9t^2$

   (i) Find the time when height $= 245$m

   (ii) Find the time it takes the rocket to hit the ground.

b. A coin, thrown upward at time $t = 0$ from an office in the Sears Tower, has height in feet above the ground t seconds later given by $h(t) = -16t^2 + 64t + 960$.

   (i) From what height is the coin thrown?

   (ii) At what time does the coin reach the ground?

Worksheet 8.2

1. A ball is thrown upward from the ground with an initial velocity of 96 ft/s, and it reaches a height h(t) after time t as described by the formula, $h(t) = 96t - 16t^2$.

    a. Find the time t when $h(t) = 80$ ft by solving the quadratic equation.

    b. Graph $h(t)$ versus from time, $t = 0$ to the time when the ball hits the ground.

    c. From the graph in b, determine the time when the ball hits the ground.

    d. Write the quadratic equation to determine the time when ball hits the ground and solve the equation.

Worksheet 8.3

1. A golfer hits a ball with an initial velocity of $v_0 = 96$ ft/s at an angle of $\theta = 50°$. If we neglect air resistance, the following equation describe the x and y positions (or the path) of the golf ball in flight as a function of time, t.

$x(t) = 61.7t$ $\qquad$ $y(t) = 73.54t - 16.1t^2$

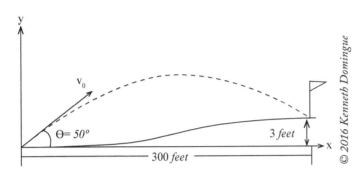

a. Find the times, t, when $y = 50$ ft, $y = 100$ ft.
b. How long does it take the golf ball to land on the green?
c. The x position of the ball when it lands on the green.
d. The maximum height of the golf ball.
e. Write the quadratic function, $y = f(x)$

Additional Warm-Up Problems

a. The average weight of a baby during the first year of life is roughly a quadratic function of time. At month m, its average weight, w, in pounds, can be approximated by $w(m) = -0.042m^2 + 1.75m + 8$.

   (i) What is the average weight of a one-year-old baby?

   (ii) At what month would the baby reaches a weight of 15 pound?

b. The stopping distance in feet, d, of a car traveling at v miles per hour (mph), is given by: $d = 2.2v + v^2/20$.

   (i) What is the stopping distance of a car traveling 60 mph? 70 mph?

   (ii) If the stopping distance of a car is 500 feet, how fast was the car traveling when it braked?

Additional Homework Problems

1. A ball is thrown from an angle, $\theta = 30°$, from a height of 10 ft at a speed of 100 ft/s and later strikes the ground as shown. The x and y positions of the ball are given as functions of time as follows:

   $x(t) = 86.6t$ (ft)

   $y(t) = 10 + 50t - 16.1t^2$ (ft)

   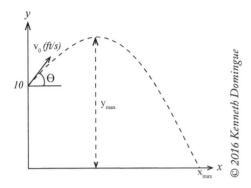
   © 2016 Kenneth Domingue

   a. Find the times when $y = 0$. Find the time for ball to reach maximum height.
   b. Find the maximum height.
   c. Find the x position when ball is at maximum height.
   d. Find x when the ball hits ground.
   e. Write the quadratic function, $y = f(x)$.

2. A lamp and a 20 ohm resistor are connected to a 120 V power supply as shown. Suppose the power, P, delivered by the power supply is described by the equation $P = I^2R + IV$, where I = current in ampere, R = resistance in ohms, and V = voltage in volts. If $P = 96$ watt (W), determine the current I. Note: 1 watt = 1 ampere-volt; recall from Ohm's law, $V = IR$, where V is in volt, I in ampere, and R in ohms.

   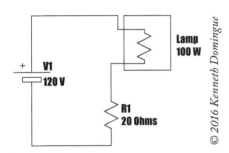
   © 2016 Kenneth Domingue

3. A rectangle has length x and width y. Suppose the area of the rectangle is 36 m^2 and the perimeter is 30 m. Write the quadratic equation that would allow you to solve for x and y. Determine x and y.

4. Suppose the radius of an oil spill increases with time such that $r(t) = 2t + 1$. The cost of clean-up in dollars is calculated by $C(t) = 800r^2 + 8800$. How fast must a team respond to the spill if the clean-up cost must be kept to less than $8,000? To less than $10,000?

Chapter 9
Exponential and Logarithm Functions

(Pressure-Volume Relations in IC Engine and Engineering Economy)

Recall an exponential function, Q, has the form:

$$Q = f(t) = a(b^t)$$

where b is called the base of the function

   t is called the exponent

   a and b are constants and $b > 0$

In this chapter, we will look at two examples of exponential function in engineering:

1. PV^n = constant. This equation describes the relationship between pressure and volume during compression and expansion of a gas in a piston-cylinder assembly of an internal combustion engine, and
2. $FV = PV(1 + i)^n$. This equation in engineering economy relates the present value (PV) of an investment or asset to its future value (FV) with interest rate per period, i, for n interest periods.

We will also learn to take logarithm of the exponential function so we can simplify the algebraic operations to take advantage of what we know about straight lines in solving these two types of engineering problems.

The Learning Outcomes for this chapter are

- Apply the appropriate exponential equation to solve problems involving pressure-volume of a gas during compression and expansion in a piston-cylinder assembly of an internal combustion engine, and engineering economy.
- Demonstrate proficiency in exponential and logarithm function rules.

9.1 Some Rules Governing Exponential Functions

Recall the following rules governing exponential functions from Algebra II that we will need in this chapter:

9.1.1 For expressions with a Common Base

$$(a^n)(a^m) = a^{n+m}$$

$$(a^m)^n = a^{mn}$$

$$\frac{a^n}{a^m} = a^{n-m}$$

9.1.2 For expressions with a Common Exponent

$$(ab)^n = a^n b^n \qquad \text{where n is an integer}$$

$$\left(\frac{a}{b}\right)^n = \frac{a^n}{b^n}$$

9.1.3 For expressions with Zero and Negative Integer Exponents

$$a^0 = 1$$

$$a^{-n} = \frac{1}{a^n}$$

9.1.4 For Expressions with Fractional Exponents

$$a^{\frac{1}{n}} = \sqrt[n]{a}$$

$$a^{\frac{m}{n}} = \sqrt[n]{a^m}$$

9.2 Properties of Logarithm Functions

Recall the following rules of logarithm functions from Algebra II that we will need in this chapter:

9.2.1 $log\ (ab) = log\ a + log\ b$

9.2.2 $\log\left(\dfrac{a}{b}\right) = \log a - \log b$

9.2.3 $\log\ (b^t) = t \log b$

9.3 Converting an Exponential Function to a Linear Function

There are many instances in which you can solve more easily a problem described mathematically by an exponential function by converting the exponential function into a linear function or equation of a straight line. This is done by taking logarithm on both sides of the equation.

Assume an exponential function

$Q = ab^t$ where a and b are constants and t is the exponent

Take logarithm on both sides of the above equation:

$\rightarrow \log Q = \log(ab^t) = \log a + t\log b$

Therefore a graph of $\log Q$ versus $\log b$ would be a straight line with slope t.

$$\text{Slope } t = \frac{\log Q_2 - \log Q_1}{\log b_2 - \log b_1}$$

9.4 Compression and Expansion of a Gas in the Piston-Cylinder Assembly of an Internal Combustion (IC) Engine

Consider Figure 9.1, which represents the position of the piston inside a piston-cylinder assembly of a 4-stroke internal combustion engine. An air-gasoline mixture is drawn into the cylinder chamber and it is compressed by the piston moving upward. Then the compressed air-gasoline mixture is ignited by an electric spark and it expands, pushing the piston downward and spinning the flywheel through the crankshaft. The spent air-gasoline mixture is exhausted.

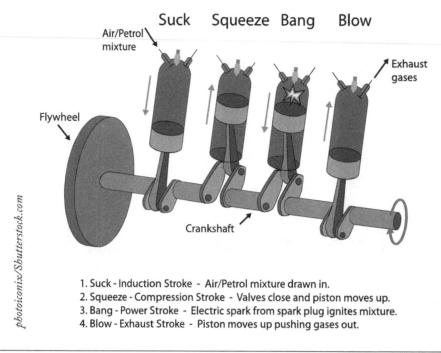

Four Stroke Engine

Suck Squeeze Bang Blow

Air/Petrol mixture

Exhaust gases

Flywheel

Crankshaft

1. Suck - Induction Stroke - Air/Petrol mixture drawn in.
2. Squeeze - Compression Stroke - Valves close and piston moves up.
3. Bang - Power Stroke - Electric spark from spark plug ignites mixture.
4. Blow - Exhaust Stroke - Piston moves up pushing gases out.

photoiconix/Shutterstock.com

Figure 9.1 A 4-stroke engine showing the compression and expansion cycle

The relationship between pressure and volume of the gas inside the piston-cylinder assembly of an internal comustion engine during compression and expansion can be described by the following exponential equation:

PV^n = constant

where P = pressure

V = volume

n = a numerical constant depending on the compression-expansion process

There are three general types of problems involving the exponential function, PV^n = constant. We will now apply algebraic operations and exponential and logarithm rules to solve each type of the problems.

9.4.1 Find P₂ or V₂, given initial conditions of (P₁, V₁) and V₂ or P₂

Start with PV^n = constant

$$P_1 V_1^{\,n} = \text{constant} = P_2 V_2^{\,n}$$

Divide both sides of equation by P_1

$$\rightarrow \left(\frac{\cancel{P_1} V_1^n}{\cancel{P_1}} \right) = \left(\frac{P_2 V_2^n}{P_1} \right)$$

Divide both sides of equation by V_2^n

$$\rightarrow \left(\frac{V_1^n}{V_2^n} \right) = \left(\frac{P_2 \cancel{V_2^n}}{P_1 \cancel{V_2^n}} \right) = \left(\frac{P_2}{P_1} \right)$$

Multiple both sides of equation by P_1

$$\rightarrow P_1 \left(\frac{V_1^n}{V_2^n} \right) = \cancel{P_1} \left(\frac{P_2}{\cancel{P_1}} \right) = P_2$$

$$\rightarrow P_2 = P_1 \left(\frac{V_1^n}{V_2^n} \right) = P_1 \left(\frac{V_1}{V_2} \right)^n$$

To obtain an equation for V_2, start with $\left(\frac{V_1^n}{V_2^n} \right) = \left(\frac{P_2}{P_1} \right)$ or $\left(\frac{P_2}{P_1} \right) = \left(\frac{V_1}{V_2} \right)^n$

Take reciprocal of both sides of equation

$$\rightarrow \left(\frac{P_1}{P_2} \right) = \left(\frac{V_2}{V_1} \right)^n$$

Raise both sides of equation to exponent $\left(\frac{1}{n} \right)$

$$\rightarrow \left(\frac{P_1}{P_2} \right)^{\frac{1}{n}} = \left(\left(\frac{V_2}{V_1} \right)^n \right)^{\frac{1}{n}} = \frac{V_2}{V_1}$$

Multiply both sides of equation by V_1

$$\rightarrow \cancel{V_1}\left(\frac{V_2}{\cancel{V_1}}\right) = V_1\left(\frac{P_1}{P_2}\right)^{\frac{1}{n}}$$

Therefore, $V_2 = V_1\left(\dfrac{P_1}{P_2}\right)^{\frac{1}{n}}$

9.4.2 Find P_1 or V_1, given final conditions (P_2, V_2) and V_1 or P_1

Again, start with $P_1V_1^n = \text{constant} = P_2V_2^n$

Divide both sides of equation by P_2

$$\rightarrow \left(\frac{P_1V_1^n}{P_2}\right) = \left(\frac{\cancel{P_2}V_2^n}{\cancel{P_2}}\right) = V_2^n$$

Divide both sides of equation by V_1^n

$$\rightarrow \left(\frac{P_1\cancel{V_1^n}}{P_2\cancel{V_1^n}}\right) = \frac{V_2^n}{V_1^n} = \left(\frac{V_2}{V_1}\right)^n$$

Multiple both sides of equation by P_2

$$\rightarrow \cancel{P_2}\left(\frac{P_1}{\cancel{P_2}}\right) = P_2\left(\frac{V_2}{V_1}\right)^n$$

Therefore, $P_1 = P_2\left(\dfrac{V_2}{V_1}\right)^n$

To obtain V_1, start with the equation $\dfrac{P_1}{P_2} = \left(\dfrac{V_2}{V_1}\right)^n$

Raise both sides of equation to exponent $\left(\dfrac{1}{n}\right)$

$$\rightarrow \left(\frac{P_1}{P_2}\right)^{\frac{1}{n}} = \left(\left(\frac{V_2}{V_1}\right)^n\right)^{\frac{1}{n}} = \left(\frac{V_2}{V_1}\right)$$

Multiply both sides of equation by V_1

$$\rightarrow V_1 \left(\frac{P_1}{P_2} \right)^{\frac{1}{n}} = \cancel{K} \left(\frac{V_2}{\cancel{K}} \right) = V_2$$

Divide both sides of equation by $\left(\dfrac{P_1}{P_2} \right)^{\frac{1}{n}}$

$$\rightarrow \frac{V_1 \cancel{\left(\dfrac{P_1}{P_2} \right)^{\frac{1}{n}}}}{\cancel{\left(\dfrac{P_1}{P_2} \right)^{\frac{1}{n}}}} = \frac{V_2}{\left(\dfrac{P_1}{P_2} \right)^{\frac{1}{n}}}$$

Therefore, $V_1 = \dfrac{V_2}{\left(\dfrac{P_1}{P_2} \right)^{\frac{1}{n}}} = V_2 \left(\dfrac{P_1}{P_2} \right)^{-\frac{1}{n}}$

9.4.3 Find n, given initial and final conditions of (P$_1$, V$_1$) and (P$_2$, V$_2$)

Start again with equation $P_1 V_1^n = \text{constant} = P_2 V_2^n$

Take logarithm of both sides of equation

$$\rightarrow \log\left(P_1 V_1^n \right) = \log\left(P_2 V_2^n \right)$$

Apply logarithm function rules

$$\rightarrow \log P_1 + n\left(\log V_1 \right) = \log P_2 + n\left(\log V_2 \right)$$

Collect the like terms (P and V) on one side of the equation

$$\rightarrow \log P_1 - \log P_2 = n\left(\log V_2 \right) - n\left(\log V_1 \right)$$

Apply logarithm function rules

$$\rightarrow \log\left(\frac{P_1}{P_2} \right) = n\left(\log\left(\frac{V_2}{V_1} \right) \right)$$

Divide both sides of equation by $\log\left(\dfrac{V_2}{V_1}\right)$

$$\rightarrow \frac{\log\left(\dfrac{P_1}{P_2}\right)}{\log\left(\dfrac{V_2}{V_1}\right)} = \frac{n\left(\log\left(\dfrac{V_2}{\cancel{V_1}}\right)\right)}{\log\left(\dfrac{V_2}{\cancel{V_1}}\right)} = n$$

Therefore, $n = \dfrac{\log\left(\dfrac{P_1}{P_2}\right)}{\log\left(\dfrac{V_2}{V_1}\right)}$ or $n = \dfrac{\log P_1 - \log P_2}{\log V_2 - \log V_1}$

9.5 Application of Exponential Function in Engineering Economy

Engineers, engineering technologists, and applied scientists need to consider engineering economy in their daily work. For example, they will need to consider return of investment in buying CNC milling stations, one of which is shown in Figure 9.2, to launch a new line of products or to improve productivity. They will need to calculate the future and current values of investing in a CNC million station and determine how much more productivity gain they can achieve with such an investment.

Dmitry Kalinovsky/Shutterstock.com

Figure 9.2 Technicians working at CNC milling machine stations

Therefore, every aspect of an engineering project involves the economics of the project. The following exponential function allows you to calculate the future-value *(FV)* of an investment of present value *(PV)* as a function of interest rate, *i*, and interest periods, *n*.

$FV = (PV) (1+i)^n$

where i is the interest rate per interest period

              n is the number of compounding periods or expected life of investment asset.

              i is known as the compound interest rate if value stays the same for each interest period.

Besides calculating the future value (FV), there are the following types of engineering economy problems associated with this exponent function:

9.5.1 Find Present Value from Future Value, Interest Rate, and Interest Periods

Start out with the exponential function, $FV = (PV) (1+i)^n$

Divide both sides of equation by PV
$$\frac{(FV)}{(PV)} = \frac{\cancel{(PV)}(1+i)^n}{\cancel{(PV)}} = (1+i)^n$$

$$\rightarrow \frac{FV}{(1+i)^n} = \frac{(PV)\cancel{(1+i)^n}}{\cancel{(1+i)^n}} = (PV)$$

Therefore, $(PV) = \dfrac{(FV)}{(1+i)^n}$

9.5.2 Find Interest Period from Present Value, Future Value, and Interest Rate

Again start out with the exponential function, $FV = (PV) (1+i)^n$

Divide both sides of equation by (PV)

$$\rightarrow \frac{FV}{PV} = \frac{(PV)(1+i)^n}{(PV)} = (1+i)^n$$

Take logarithm of both sides of equation

$$\rightarrow \log\left(\frac{FV}{PV}\right) = \log\left((1+i)^n\right) = n\left(\log(1+i)\right)$$

Divide both sides of equation by $n(\log(1+i))$

$$\rightarrow \frac{\log\left(\dfrac{FV}{PV}\right)}{\log(1+i)} = \frac{n\left(\cancel{\log(1+i)}\right)}{\cancel{\log(1+i)}} = n$$

Therefore, $n = \dfrac{\log\left(\dfrac{FV}{PV}\right)}{\log(1+i)}$

9.5.3 Find Interest Rate from Present Value, Future Value, and Interest Period

Again start out with the exponential function, $FV = (PV)(1+i)^n$

Divide both sides of equation by (PV)

$$\rightarrow \frac{FV}{PV} = \frac{(PV)(1+i)^n}{(PV)} = (1+i)^n$$

Raise both sides of equation to exponent $\dfrac{1}{n}$

$$\rightarrow \left(\frac{FV}{PV}\right)^{\frac{1}{n}} = \left((1+i)^n\right)^{\frac{1}{n}} = 1+i$$

Subtract "1" from both sides of equation

$$\rightarrow \left(\frac{FV}{PV}\right)^{\frac{1}{n}} - 1 = (1+i) - 1 = i$$

Therefore, $i = \left(\dfrac{FV}{PV}\right)^{\frac{1}{n}} - 1 = \left(\dfrac{PV}{FV}\right)^n - 1$

9.6 Example Problems

9.6.1 Example Problem #1

A frictionless piston-cylinder device contains nitrogen (N_2) at an initial pressure of 100 kPa and a volume of 1.8 m^3. The nitrogen gas is then compressed slowly until the pressure is double. It is known during the compression process, the relationship between pressure and volume of the gas can be described by the equation $PV^{1.4}$ = constant. Calculate the final volume.

Overview of Solution

In this problem, you are given the initial conditions of P_1 and V_1, as well as final pressure, P_2 (double P_1). You are asked to determine V_2. You can apply the equation $P_1V_1^n$ = constant = $P_2V_2^n$ as well as taking logarithm of PV^n to solve for V_2.

Start with $P_1V_1^n$ = constant = $P_2V_2^n$

Divide both sides of equations by V_2^n, and then divide both sides of equation by P_1

$$\left(\frac{V_1^n}{V_2^n}\right)=\left(\frac{P_2}{P_1}\right) \text{ or } \left(\frac{P_2}{P_1}\right)=\left(\frac{V_1}{V_2}\right)^n$$

Raise both sides of equation to exponent $\left(\dfrac{1}{n}\right)$

$$\rightarrow\left(\frac{P_2}{P_1}\right)^{\frac{1}{n}}=\left(\left(\frac{V_1}{V_2}\right)^n\right)^{\frac{1}{n}}=\frac{V_1}{V_2}$$

Take the reciprocal of both sides of equation

$$\rightarrow\frac{V_2}{V_1}=\left(\frac{P_1}{P_2}\right)^n$$

Multiple both sides of equation by V_1

$$\rightarrow(\cancel{V_1})\left(\frac{V_2}{\cancel{V_1}}\right)=(V_1)\left(\frac{P_1}{P_2}\right)^n$$

Therefore, $V_2=V_1\left(\dfrac{P_1}{P_2}\right)^{\frac{1}{n}}$

Substitute the values of P_1 = 100 kPa, V_1 = 1.8 m^3; P_2 = 2P_1 = 200 kPa, and n =1.4 into the above equation

$$V_2=\left(1.8m^3\right)\left(\frac{100kPa}{200kPa}\right)^{\frac{1}{1.4}}=1.097m^3$$

9.6.2 Example Problem #2

What is the value of an initial deposit that would yield $15,000 in ten years at a compound interest rate of 3.5%?

Overview of Solution

In this problem, you are given the compound interest rate, i, and the future value, FV, in ten years, n, and you are asked to determine the initial value, PV. Here, 3.5% interest rate means $i = 0.035$.

Start with equation $FV = (PV)(1+i)^n$

$$\rightarrow PV = \frac{FV}{(1+i)^n} = \frac{\$15{,}000}{(1+0.035)^{10}} = \$10{,}633.78$$

You will need to make an initial deposit of $10,633.78 to yield $15,000 in 10 years at compound interest rate of 3.5%.

Worksheet 9.1

a. Assume you make a $1,000 deposit into a saving account that has a compound interest rate of 5% per year. What would be its value after 20 years?

b. What is the value of an initial deposit that would yield $15,000 in 10 years at a compound interest rate of 3.5%?

Worksheet 9.2

1. One ft^3 of Nitrogen is compressed in a reversible process in a cylinder from 14.7 psi (pounds per square inch) and 60F to 60 psi. During the compression process, the relation between pressure and volume is $PV^{1.3}$ = constant. What is the volume at the end of the compression process?

Worksheet 9.3

2. An air-standard diesel cycle has a compression ratio of 15. At the beginning of the compression process, the pressure is 14.7 psi (pounds per square in). If the relation between pressure and volume is given by the expression PV^k = constant. If $k = 1.4$, what is the pressure at the end of the compression process?

Additional Warm-Up Problems

a. The pressure and volume of a gas undergoing a reversible process in a piston-cylinder assembly can be described by the equation PV^n = constant. If P = 20 psi (pounds per square inch) and V = 1 ft³, n = 1.4, what is the numerical value and units of the constant.

b. The relation between pressure and volume in an expansion/contraction process is governed by the equation, PV = constant. If P_1 = 20 psi, V_1 = 1 ft³, what is P_2 if V_2 = 3 ft³?

Additional Homework Problems

1. $0.5\ m^3$ of helium is expanded in a reversible process in a cylinder from 1.5 atmosphere to a volume of $1.25\ m^3$ at which time the pressure is 0.5 atmosphere. During the expansion process, the relationship between pressure and volume is PV^n = constant. What is n?

2. An asset worth $10,000 becomes $15,276 at a compound interest rate of 4.5%. How many years would the current asset need to grow to attain the future value?

3. What annual compound interest rate would double the value of an investment after 15 years?

4. A nitrogen gas is compressed reversibly from 14.7 psi (pounds per square inches), $1\ ft^3$ to $0.25\ ft^3$. During the compression process, the relationship between pressure and volume of the nitrogen gas is given by PV = constant.

   (a) What is the final pressure?

   (b) Graph pressure (vertical axis) versus volume (horizontal axis) of nitrogen during the compression process, in increment volume change of $0.125\ ft^3$.

5. The annual compound interest rate for an investment is 2.5%. How long do you need to hold the investment to double its value?

Chapter 10
Natural Exponential and Natural Logarithm Functions

(Diffusion Coefficient and Light Absorption)

10.1 Natural Exponential Function

Recall that $Q = ae^{bt}$ is called a natural exponential function (or exponential function with base e) where a and b are constants

> e is known as base e and has a numerical value 2.71828…
>
> t is in the independent variable or the input
>
> Q is the dependent variable or the output

10.1 Exponential Growth and Decay Functions

For positive values of b *(+b)*, Q describes exponential growth rates. See Figure 10.1. For negative values of b *(–b)*, Q describes exponential decay rates. See Figure 10.2.

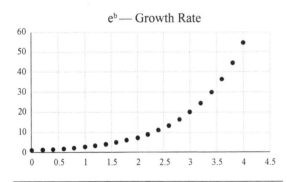

Figure 10.1 Exponential Growth Function

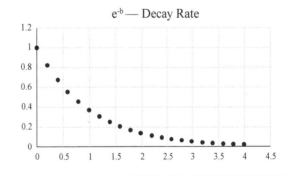

Figure 10.2 Exponential Decay Function

10.1.1 Properties of Natural Exponent Function

10.1.1.1 $e^a \cdot e^b = e^{a+b}$

10.1.1.2 $(e^a)^b = e^{ab}$

10.1.1.3 $\left(\dfrac{e^a}{e^b}\right)^n = e^{an-bn}$

10.1.1.4 $e^0 = 1$

10.1.1.5 $e^{-a} = \dfrac{1}{e^a}$

10.2 Natural Logarithm Function

The natural logarithm of a number is its logarithm to base e[1].

If $y = \ln x \rightarrow e^y = x$

Or

If $e^y = x$, then $y = \ln x$

Furthermore, $\ln e = 1$

10.2.1 Properties of Natural Logarithm

The following equations describe the natural logarithm's properties

10.2.1.1 $\ln(e^N) = N$ for all N

10.2.1.2 $e^{\ln N} = N$ for $N > 0$

10.2.1.3 $\ln(ab) = \ln a + \ln b$

10.2.1.4 $\ln\left(\dfrac{a}{b}\right) = \ln a - \ln b$

10.2.1.5 $\ln(b^t) = t\,(\ln b)$

---

[1] McCallum, W.G., Connally, E., Hughes-Hallett, D., et al (2015). *Algebra: Form and Function, Second Edition.* Hoboken, NY: Wiley.

10.3 Converting a Natural Exponential Function to a Linear Function

There are many instances in which you can more easily solve a problem described mathematically by a natural exponential function by converting the natural exponential function into a linear function or equation of a straight line. This can be accomplished by taking natural logarithm of both sides of the equation.

Assume a natural exponential function

$$y = ae^{bx}$$

where a and b constants,

e is known as base e and a numerical value of 2.71828…

x is in the input or independent variable

y is the output or dependent variable

Take natural logarithm on both sides of equation

$$\rightarrow ln\ (y) = ln\ (ae^{bx}) = ln\ a + ln\ (e^{bx}) = ln\ a + bx\ ln\ (e) = ln\ a + bx$$

Therefore, a graph of $ln\ (y)$ versus x will yield a straight line with slope b and intercept $ln\ a$.

Slope, $b = \dfrac{\Delta \ln y}{\Delta x} = \dfrac{(\ln y_2 - \ln y_1)}{(x_2 - x_1)}$

10.4 An Example of an Exponential Decay Function: Diffusion

Diffusion describes the process of mass transport in a solid by atomic motion of the diffusing atoms. Diffusion is very important in the fabrication of semiconducting devices such as p-n junctions and Integrated Circuits (IC). A diffusion process called surface hardening has been used to increase the surface toughness of metals, ball bearings, and cutting tools.

Figure 10.3 A technician holding a silicon wafer

Figure 10.3 above shows a technician holding a silicon chip with circuits etched on the wafer that are created with processes such as diffusion.

Diffusion coefficient, D, is an indicator of the rate of atomic diffusion:

$$D = D_o e^{\frac{-Q}{RT}}$$

where D is the diffusion coefficient at temperature T

           Q is the activation energy for diffusion (in Joule/mole).

           R is the Universal Gas constant and has a value of 8.31 *Joule/mole-°K*

           D_o is a temperature-independent constant (in *m²/s*)

           T is temperature (in absolute temperature scale, in *°K*)

As you can see from the expression, the value of D decreases as T increases.

In the above, Q is a material property that is a function of the host atoms as well as the diffusing atom.

The activation energies for diffusion, Q, of different types of diffusing atoms in various hosts can be found in engineering handbooks.

There are four (4) common types of problems involving the diffusion coefficient.

10.4.1 Find D given the values D_o, Q and T

Since $D = D_o e^{\frac{-Q}{RT}}$

First, convert temperature T to absolute temperature scale:

$$T\,(°K) = T\,(°C) + 273$$

Recall that $R = 8.31$ Joule/mole-°K

Substitute the values of D_o, Q and T into equation and solve for D.

10.4.2 Given D, D_o, and Q. Find T.

Again start out with $D = D_o e^{\frac{-Q}{RT}}$

Divide both sides of equation by D_o

$$\rightarrow \frac{D}{D_0} = \frac{D_o e^{\frac{-Q}{RT}}}{D_o} = e^{\frac{-Q}{RT}}$$

Take natural logarithm on both sides of equation and recall ($ln\ e = 1$)

$$\rightarrow \ln\left(\frac{D}{D_o}\right) = \ln(e^{\frac{-Q}{RT}}) = \frac{-Q}{RT}\underset{=1}{\ln e} = \frac{-Q}{RT}$$

Multiple both sides of equation by T

$$\rightarrow \ln\left(\frac{D}{D_o}\right)(T) = \left(\frac{-Q}{R\cancel{T}}\right)(\cancel{T}) = \frac{-Q}{R}$$

Divide both sides of equation by $\ln\left(\dfrac{D}{D_o}\right)$

$$\rightarrow \frac{\cancel{\ln\left(\dfrac{D}{D_o}\right)}T}{\cancel{\ln\left(\dfrac{D}{D_o}\right)}} = \frac{\dfrac{-Q}{R}}{\ln\left(\dfrac{D}{D_o}\right)}$$

Therefore, $T = \dfrac{\left(\dfrac{-Q}{R}\right)}{\ln\left(\dfrac{D}{D_o}\right)}$

We realize $\dfrac{1}{\ln\left(\dfrac{D}{D_o}\right)} = \ln\left(\dfrac{D_o}{D}\right)$ on right-hand side equation

$$T(^{\circ}K) = \left(\frac{-Q}{R}\right)\ln\left(\frac{D_o}{D}\right)$$

Therefore, $T(^{\circ}C) = T(^{\circ}K) - 273$

10.4.3 Find Q from (D$_1$, T$_1$) and (D$_2$ and T$_1$)

Again start with definition of Diffusion Coefficient at the two given conditions (D_1, T_1) and (D_2 and T_1)

$$D_1 = D_o e^{\frac{-Q}{RT_1}} \quad \text{and}$$

$$D_2 = D_o e^{\frac{-Q}{RT_1}}$$

Divide equation for D_1 by equation for D_2

$$\frac{D_1}{D_2} = \frac{D_o e^{\frac{-Q}{RT_1}}}{D_o e^{\frac{-Q}{RT_2}}} = \frac{e^{\frac{-Q}{RT_1}}}{e^{\frac{-Q}{RT_2}}} = \frac{e^{\frac{Q}{RT_2}}}{e^{\frac{Q}{RT_1}}}$$

Take natural logarithm of both sides of above equation

$$\rightarrow \ln\left(\frac{D_1}{D_2}\right) = \ln\left(\frac{e^{\frac{Q}{RT_2}}}{e^{\frac{Q}{RT}}}\right) = \frac{Q}{RT_2} - \frac{Q}{RT_1} = \frac{Q}{R}\left(\frac{1}{T_2} - \frac{1}{T_1}\right)$$

In the above equation, $\ln\left(\frac{D_1}{D_2}\right) = \ln D_1 - \ln D_2$ can be think of as "rise" or change (Δ) in $\ln D$ in a graph of $\ln D$ versus $\frac{1}{T}$.

The term, $\left(\frac{1}{T_2} - \frac{1}{T_1}\right)$ can be think of as "run" or change in $\frac{1}{T}$ in a graph of $\ln D$ versus $\frac{1}{T}$.

The slope of such a graph, $\frac{rise}{run}$, or change in $\ln D$ per change in $\frac{1}{T}$, is equal to $\frac{-Q}{R}$

Therefore, $Q = -R(\text{slope value}) = R = \dfrac{\ln\left(\dfrac{D_1}{D_2}\right)}{\left(\dfrac{1}{T_1} - \dfrac{1}{T_2}\right)}$

10.4.4 Find D(T$_3$) from (D$_1$, T$_1$) and (D$_2$, T$_2$)

Again, start with equation for Diffusion coefficient $D(T)$

$$D = D_o e^{\frac{-Q}{RT}}$$

Take natural logarithm on both sides of equation

$$\ln D = \ln D_o - \frac{Q}{R}\left(\frac{1}{T}\right)$$

Write the above equation three times for (D_1, T_1), (D_2, T_2) and (D_3, T_3)

$$\ln D_1 = \ln D_o - \frac{Q}{R}\left(\frac{1}{T_1}\right)$$

$$\ln D_2 = \ln D_o - \frac{Q}{R}\left(\frac{1}{T_2}\right)$$

$$\ln D_3 = \ln D_o - \frac{Q}{R}\left(\frac{1}{T_3}\right)$$

Subtract middle equation from top equation

$$\rightarrow \ln D_1 - \ln D_2 = \ln\left(\frac{D_1}{D_2}\right) = \left(\frac{Q}{R}\right)\left(\frac{1}{T_2} - \frac{1}{T_1}\right)$$

Divide both sides of equation by $\left(\dfrac{1}{T_2} - \dfrac{1}{T_1}\right)$

$$\left(\frac{Q}{R}\right) = \frac{\ln\left(\dfrac{D_1}{D_2}\right)}{\left(\dfrac{1}{T_2} - \dfrac{1}{T_1}\right)}$$

Subtract bottom equation from top equation

$$\rightarrow \ln D_1 - \ln D_3 = \ln\left(\frac{D_1}{D_3}\right) = \left(\frac{Q}{R}\right)\left(\frac{1}{T_3} - \frac{1}{T_1}\right)$$

Divide both sides of equation by $\left(\dfrac{1}{T_3} - \dfrac{1}{T_1}\right)$

$$\left(\frac{Q}{R}\right) = \frac{\ln\left(\dfrac{D_1}{D_2}\right)}{\left(\dfrac{1}{T_2} - \dfrac{1}{T_1}\right)}$$

Equating the two expressions for $\left(\dfrac{Q}{R}\right)$

$$\left(\frac{Q}{R}\right) = \frac{\ln\left(\dfrac{D_1}{D_2}\right)}{\left(\dfrac{1}{T_2} - \dfrac{1}{T_1}\right)} = \frac{\ln\left(\dfrac{D_1}{D_3}\right)}{\left(\dfrac{1}{T_3} - \dfrac{1}{T_1}\right)}$$

Multiple both sides of equation by $\left(\dfrac{1}{T_3} - \dfrac{1}{T_1}\right)$

$$\frac{\ln\left(\dfrac{D_1}{D_2}\right)\left(\dfrac{1}{T_3} - \dfrac{1}{T_1}\right)}{\left(\dfrac{1}{T_2} - \dfrac{1}{T_1}\right)} = \ln\left(\frac{D_1}{D_3}\right) = \ln D_1 - \ln D_3$$

Switch the term $\dfrac{\ln\left(\dfrac{D_1}{D_2}\right)\left(\dfrac{1}{T_3} - \dfrac{1}{T_1}\right)}{\left(\dfrac{1}{T_2} - \dfrac{1}{T_1}\right)}$ from left hand-side to right-hand side of the equation

Switch the term $\ln D_3$ from right-hand side to left-hand side of equation

$$\ln D_3 = \ln D_1 - \frac{\ln\left(\dfrac{D_1}{D_2}\right)\left(\dfrac{1}{T_3} - \dfrac{1}{T_1}\right)}{\left(\dfrac{1}{T_2} - \dfrac{1}{T_1}\right)} = A$$

$$D_3 = e^A$$

10.6 Another Engineering Example Described by Natural Exponential/Logarithm Functions— Light Absorption and Phosphorescence

10.6.1 Light Absorption

The potential of solar energy in meeting a significant portion of the total energy need in the United States is being realized. Picture 10.4 below shows a solar cell found on solar panels.

neijia/Shutterstock.com

Figure 10.4. A hand holding part of a broken piece of solar cell

The intensity of light can be absorbed by passing it through light absorbing materials. Light absorption in a material is characterized by the material property, *absorption coefficient*, which gives the light's intensity as a function of distance traveled by light through the material

$$I = I_o e^{-\mu x}$$

where I_o = initial intensity

           x = distance traversed by light through the material (units m)

           μ = absorption coefficient

10.6.2 Phosphorescence

Figure 10.5 shows the phenomenon phosphorescence causing the blue cat eyes to glow in the dark.

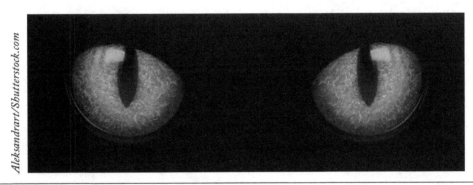

Figure 10.5 Blue cat eyes glowing in dark

Phosphorescence describes the phenomenon in special materials, called phosphorescent materials, to give off radiation including light when the material is stimulated externally to cause the light emission.

$$I = I_o e^{\frac{-t}{\tau}}$$

Where I_o = initial intensity

           t = time following removal of external stimulus causing phosphorescence

           τ = relaxation time

10.7 Example Problems

10.7.1 Example Problem #1

At what temperature will the diffusion coefficient for diffusion of zinc in copper have a value of 2.6×10^{-16} m^2/s, if given an activation energy for diffusion, Q, of 189,000 Joule/mole, and D_o of 2.4×10^{-5} m^2/s.

Overview of Solution

Since $D = D_o e^{\frac{-Q}{RT}}$ and we are asked to find T, which is the exponent, we want to take natural logarithm of the exponential function to simply the computation.

Rewrite equation for diffusion coefficient as $\dfrac{D}{D_o} = e^{\frac{-Q}{RT}}$

Take natural logarithm on both sides of the equation

$$\rightarrow \ln\left(\frac{D}{D_o}\right) = \ln\left(e^{\frac{-Q}{RT}}\right) = \left(\frac{-Q}{R}\right)\frac{1}{T}$$

Multiple both sides of above equation by T

$$\rightarrow \ln\left(\frac{D}{D_o}\right)(T) = \left(\frac{-Q}{R}\right)\frac{1}{\cancel{T}}\,(\cancel{T}) = \left(\frac{-Q}{R}\right)$$

Divide both sides of above equation by $\ln\left(\dfrac{D}{D_o}\right)$

$$\rightarrow \frac{\ln\left(\cancel{\frac{D}{D_o}}\right)T}{\ln\left(\cancel{\frac{D}{D_o}}\right)} = \frac{\left(\dfrac{-Q}{R}\right)}{\ln\left(\dfrac{D}{D_o}\right)}$$

Therefore, $T = \dfrac{\left(\dfrac{-Q}{R}\right)}{\ln\left(\dfrac{D}{D_o}\right)} = \dfrac{\left(\dfrac{-189{,}000\,\frac{Joule}{Mole}}{8.31\,\frac{Joule}{Mole-K}}\right)}{\ln\left(\dfrac{2.6x10^{-16}\,\frac{m^2}{s}}{2.4x10^{-5}\,\frac{m^2}{s}}\right)}$

$$T = 722.3°K = (722.3 - 273)\,°C = 449.3\,°C$$

10.7.2 Example Problem #2

The intensity of a light beam after passing through a material is given by $I = I_o e^{-\mu x}$ where I_o is the initial intensity, μ is the linear absorption coefficient, and x is the distance traveled by light beam. Suppose the material is 1 mm thick and has a linear absorption coefficient of 1,000 cm^{-1}. What is the intensity of light beam after it travels through the material?

Overview of Solution

This is a straightforward problem applying the equation for the intensity of light after it travels through a material of length x and with linear absorption coefficient μ: It is best to express the final intensity as a fraction of the initial intensity, that is, $\dfrac{I}{I_o} = e^{-\mu x}$

Therefore, $\dfrac{I}{I_o} = e^{-\mu x} = e^{(-1000\ cm^{-1})(0.1\ cm)} = e^{-100} = 3.72 \times 10^{-44}$

Worksheet 10.1

a. The diffusion coefficient of nitrogen (N) in iron (Fe) can be expressed as a function of temperature by the following expression: $D = D_o \exp(-Q/RT)$ where D_o is a constant; Q is the activation energy, and R is the gas constant and has a value of 8.31 Joule/mole-K, T is the absolute temperature (K) where

$$^\circ K = {}^\circ C + 273$$

Given the activation energy of nitrogen in body-centered cubic iron is 76,570 Joule/mole, and the constant D_o has a value of 0.0047 cm²/sec, what is the diffusion coefficient at 700°C?

b. Phosphorescence is the phenomenon whereby a material continues to emit light for a length of time even after the stimulus for light emission has been stopped. For phosphorescent materials, the intensity of light emission as a function of time is given by the equation (after the original stimulus is removed)

$$I(t) = I_o e^{(-t/\tau)}$$ where I_o is the initial intensity and τ is the relaxation time.

For $CaWO_4$, it has a relaxation time of 4×10^{-6} second, determine the time in which the intensity of this phosphorescent material is deceased to 1% of the original intensity after the stimulus is removed.

Worksheet 10.2

c. The diffusion coefficient of oxygen (O) in chromium Oxide (Cr_2O_3) can be expressed as a function of temperature by the following expression: $D = D_o \exp(-Q/RT)$ where D_o is a constant; Q is the activation energy, and R is the gas constant and has a value of 9.31 Joule/mole-K, T is the absolute temperature (K)

where $°K = °C + 273$

Given the diffusion coefficient has a value of 4×10^{-15} cm²/s at 1150°C and 6×10^{-11} cm²/s at 1715°C, calculate the activation energy, Q, and the constant, D_o.

Worksheet 10.3

d. The intensity of a beam of light after passing through a material is given by $I = I_o e^{(-\mu x)}$, where I_o is the initial intensity, μ is linear absorption coefficient, and x is the distance traveled by beam of light (or thickness of the material). A material has a linear absorption coefficient of 591 cm^{-1}. Determine the thickness of the material required to absorb 50% of light traveling through the material.

Additional Homework Problems

1. The diffusion coefficients for nickel in iron are given at two temperatures:

   $D(1200°C) = 2.2 \times 10^{-15}$ m2/s   and $D(1400°C) = 4.8 \times 10^{-14}$ m²/s

   What is the value of the diffusion coefficient at $1300°C$?

2. The intensity of a phosphorescent material is reduced to 90% of its original intensity after 1.95×10^{-7} second. Determine the time required for the intensity to decrease to 1% of its original intensity. Given: $I(t) = I_o e^{(-t/\tau)}$

Appendix I
Instructions on Doing Engineering Homework

It is important for first-year college students to develop the proper method, procedure, and mindset in solving problems in engineering, engineering technology or applied sciences. A good place to start the development is bydoing homework or examination. By presenting your work in a logical and systematic fashion, it makes it easier for your instructor to provide you with feedback on your homework and, in the case of an examination, give you partial credits for your work. Later in your academic career, when you are in an internship, presenting your work in a logical and systematic method will also help your engineering supervisor to spot mistakes and give you feedback.

Here are the steps you should adopt in doing homework in engineering:

1. Always use engineering paper for your homework.

2. Start each problem on a new page and write down, at the top of the page, your name, the course (title and number) and the homework problem number.

3. Show all the steps; if you skip steps and give only the answer, the instructor or the grader of homework will not be able to provide feedback or to give you partial credits in case you made a mistake. Also, when you look back to examine your own work, you will not know what you did to solve the problem. Showing all the steps in your work is a good habit to develop now for your future career as an engineer-in-training, because it will help your supervisor monitoring your work to spot mistakes and to provide feedback.

4. Show the steps by working from top to bottom of the page, from left to right, line by line and begin most lines with an "=" sign on the left.

5. Include a hand-sketch to illustrate the problem you are trying to solve, when appropriate. Write down what you are asked to find and what are given.

6. Solve each step of the problem by using algebraic symbols all the way until the last step (instead of determining a numerical value for each step of the computation; otherwise, you are doing arithmetic and not algebra).

7. If the problem asks for a numerical value, plug in the given engineering parameters in the last step, and include the appropriate units.

8. If the homework requires more than one page, number the pages.

9. While your instructor may encourage students to work in teams, always do your own work. The final product you submit must be your own work.